Sudoku

Sudoko starts with a 9x9 grid that is divided into nine 3x3 subgrids. Some cells are pre-filled with numbers. Use numbers 1 through 9 to fill in the empty cells. Each row must contain all numbers from 1 to 9 with no repetition. Each column must contain all numbers from 1 to 9 with no repetition.

HOW TO PLAY

1. Identify the numbers already given in the puzzle. Focus on completing rows, columns, and subgrids with missing numbers.
2. Look for rows, columns, or subgrids with more given numbers. Eliminate possible options for empty cells based on the given numbers in the same row, column, or subgrid.
3. Scan rows, columns, and subgrids for missing numbers. Use the process of elimination to determine the correct placement of numbers.
4. Identify cells where only one number can fit based on the given numbers and the rules. Fill in those obvious numbers.
5. Continue filling in numbers based on the rules and the numbers already placed. Check for conflicts and correct any mistakes.
6. Approach the puzzle systematically, focusing on one row, column, or subgrid at a time.
7. Sudoku requires logical thinking and patience. Take your time and avoid guessing.
8. Periodically check your progress to ensure there are no conflicts or mistakes. Keep solving and filling in numbers until the entire grid is complete.

Tips:
- Pay attention to the interactions between rows, columns, and subgrids.
- Start with easier puzzles and gradually progress to more difficult ones.

Remember, Sudoku puzzles have unique solutions, and they are designed to be solvable using logical deduction. Enjoy the challenge!

All solutions are included at the back of the book.

Easy 1

```
1 . . | . 8 . | 5 . 9
. 6 2 | 3 5 . | 7 . .
. . 3 | 7 . 1 | . . 2
------+-------+------
. . . | 5 2 . | . . .
7 . . | 1 . . | . 5 .
. . . | . . . | 3 . .
------+-------+------
. . 6 | 2 . 7 | . . .
3 . 7 | 8 . 6 | 9 . .
. . 9 | . . . | . . .
```

Easy 2

```
4 . . | . 9 . | . 1 5
. . . | . 8 3 | . . .
. 9 6 | 7 . . | . . .
------+-------+------
. . . | 9 . 5 | . 6 4
. . . | . . . | 5 . .
. . 7 | . . . | . 8 .
------+-------+------
. . . | 6 8 . | 3 . .
. . . | 5 . 4 | . . 6
. 4 . | . 3 . | . . .
```

Easy 3

```
4 7 . | 9 . . | 5 . .
. 5 . | 6 . . | . . .
. . . | . 8 . | . 4 .
------+-------+------
8 . . | . 7 1 | . . .
. . . | . . . | . 9 .
. . . | . 9 . | . 6 3
------+-------+------
. . 3 | . . . | 7 . .
. . . | . . . | . 5 .
. 6 4 | 5 . . | . . 8
```

Easy 4

```
2 . . | . . 3 | . . .
3 . . | . . . | 2 . 7
. 9 . | . . 7 | . 4 .
------+-------+------
. 2 1 | . . . | 9 . .
. . . | 4 . . | 3 . .
6 . 9 | 1 . . | . 2 .
------+-------+------
. . . | . . . | 6 . .
. . 8 | . . . | 4 7 9
1 7 . | . . . | . . 8
```

Easy 5

```
. . . | . . 8 | . . .
. . . | 3 5 . | . 1 .
8 5 . | . . . | 6 7 .
------+-------+------
7 . . | . . 2 | . . .
. . 1 | . . . | 8 6 .
. . . | . . . | 2 . 7
------+-------+------
3 8 . | 4 . . | . . .
. . . | 6 1 . | 3 . 2
. . . | . . . | . . 5
```

Easy 6

```
. . 3 | 1 9 . | . . .
. . . | 2 6 8 | . . 4
. . . | . . . | . 6 .
------+-------+------
9 . 6 | . . 4 | . 2 3
. 2 . | . . . | . . .
5 3 . | . . . | 9 . .
------+-------+------
4 . . | . 1 3 | 2 . .
. . . | 8 4 . | . . .
1 . . | . . . | 3 . .
```

Easy 7

6	4			7	5			3
5				8	2	4		
						8		
				5	8	2		4
8			9			6	5	
	3	2				7		9
							6	2
	7							

Easy 8

					7		8	
9							2	
		2	5		8	1		
	6		1			5	7	
3	9			5			1	
				6				
			7				5	
7						2		
	5			3		6	9	

Easy 9

	1				4			
2		8						
		6				5		
7			6		1	9		
	6		5		9	4		1
						7		
			1					9
		4		6			2	5
8				4	7			

Easy 10

5	6				7		3	2
							4	
		8						
		3		8		2	5	
7					5	4		
							9	3
4	2		9	5		7	6	
		7						
9					2	5		

Easy 11

			3		4	8	6	
6		2						5
	3		8			9		
2			6		8		9	
4				2		7		
9								6
		9		3				
								7
				7		3	4	

Easy 12

			2				3	
		8						7
3	9						5	
			5		6	7	2	
7			9		4			3
8	5		3					
9			4					6
	6						8	2
	2			4				

Easy 13

					1			
				2			8	
		2	6	4	9			
	4					6	2	
	7		5				4	
6				7		3		9
	1	9	7			4		
	6	7		9		2		

Easy 14

		2			5	6	7	
	5							
3		8			7	4		5
5	2		1					
		4	3	8				7
		7		5				4
6			8	2				
						1		
				6				

Easy 15

			3					
9	1						3	
	5			4	2	8		
		6				9		
		2	1					
					8			
			6			5		
		1	4				2	
6	3	9		5		4	8	

Easy 16

								6
	1	4			2	8		
	8				7			2
	2	1		6	3	9		
	3						1	
4			8					
7	6	3						
8		2						7
	4					6		

Easy 17

			2	6				
				8				
	3							
7	1			3		4		
	5	8		1		3	2	
		6				7		5
1	4		2	6				7
	8			7				3
			8		1	4		

Easy 18

1	5		6					4
			5					
8				4	2			
						6		1
4	2		3					5
				5		9	4	
	4						2	
	3	5		6	8			
	6		4	3				

Easy 19

			3		1	6		
		3						
6				8	2		9	
7		4			9	2		
	2			1	6	3		
9							4	1
								8
		9						2
3			8	6				

Easy 20

	5		6					1
	1		4					3
9		6						
1		9				8	2	
	3	8				9		
5				6				
				2		4		
	6			4				8
3	8			5				

Easy 21

			5	8	3	6		
			1			5	8	
				6				
3	2					4		8
	1				8	2	7	
		4		3			6	
5								
		3		1	7			
	8		4		6			

Easy 22

		7			5			
		6	1	9	7	8		
4	8							
		8				1	4	
	9			6	8			7
7								
		1		8				
3		4	5		1			
	5				3			

Easy 23

	3		8	6		7		
			9					6
7		9				2	8	
		2						
3		7			4			
	8		4		2	9		
			3					1
			7		1		4	

Easy 24

								6
	5	1		2				
	7		5		1	8		
7	1							
5			2		7			
	6	2						9
				8	3			
8			6			1		
	2			3				7

Easy 25

	9		6		4	1		8
6	5				7			
1				8				
	4	9		7				
	2		1					
				2		8		7
		5			6		1	
	7					4	2	
9								

Easy 26

					7			
	6	4	9					
					3			
7		1						4
				7			6	
					4	8		1
1		3		6			4	9
8			4					5
4		7	3		9			6

Easy 27

					7			
		1		4				7
	4		2	5			9	
	6	9			5			
	7			1				
			7	8				
9			5	6				
		6			2	8		
	2	5	9		1	4		

Easy 28

		5					1	
6								
	2	3						
	9		1					
		1	5	3				2
	6	4			2			7
			7	3				
		9		2	4		6	
			7		1		5	

Easy 29

4		1	8					6
		2	6			7		
				2		5		4
8				7				
	6	4		5		2		
	2	7			9	1		
					6			
5								2
		8			1		5	

Easy 30

						4	7	5
	5							
9						2		
	7			5		3		
4			6					
		6	7					2
3				4		5		9
			7		6	1		4
2	4		9					

Easy 31

5			6	4				
8			5					1
		6	3	7	1			
	4							9
			4	6			7	
				3				8
7								5
6						8		
9		1	8					

Easy 32

		1	8			2	5	
	9	3			2			7
2				1	9			
		2	4				8	1
3				7				
5	7		1	9			2	
			3					
				5		8	9	

Easy 33

6								8
			4		2	7		
						2		9
			4	7		1		
			6			9		2
		9						3
	3	8		2	1			6
		2		8				

Easy 34

5		8						
	2			7	9			
3				1				
			1			2	7	
		2		3				
	3		9				8	
2				8			5	
		1					6	
	8	5			2	1		9

Easy 35

	2							7
				8			2	
				7			9	
			6					9
				9		5	8	
	7				2		1	
8								1
5	9			1	4		7	8
	6	4				9		

Easy 36

8					9			
	3	6		8				
				7				
2	9					6		
								7
	5	4		3	8		2	
9	8						5	
		3		9	2			
				5		7	8	

Easy 37

		1			8	3		
								5
3						8		2
		2	3					4
				9		2	6	3
		5					8	
9				5				6
	5	4	6			9		
	6			1	3			

Easy 38

8		7						
9					8		4	7
	4					1	2	
6					3		9	4
	7					6		
			4					3
7			6				1	
	1				9	4		
	9						8	2

Easy 39

4					2			
		5			7			6
	2	6						
			5	7	9			8
							3	5
		9	4					
3	5		8			6		
	8		6			2		
						4		

Easy 40

7	5		4					
		9	2		6	1		
						4		9
	7			9		5		2
8						7		
	2	6	8	5		9		1
						2		
								5
6	1							

Easy 41

	8		1			6		
		3						1
			3		2			7
4			2		7			8
9						7		
		1		3	9			2
				4			9	6
	7							

Easy 42

6		1			3			
	9	4						
			9	4				
9				3	4			7
		5		9				
3			6			1	5	
	6				1			5
7						3	1	
			8		7			

Easy 43

				4			2	
			5			4	1	
			7				8	
					1			
8	3				5	6	4	
			3	8				
	5		4	2				
		7		6		8		
	2	6	8			7	3	

Easy 44

8						1		2
					1	9	3	8
			5				4	
	3		6		4		8	
			3			6		
	4					5	1	
9	2							
3				6				
1				2	8			5

Easy 45

		1					6	
7			6	8				
2		4				1		
8	7			1	3			
		3				6		
	9							
		7			6	3	1	
	2					8		4
	1			9	2		7	

Easy 46

	9			5	2		4	
3			1	8				9
	8				3	2		
2		7		4		1		
1		4				5		
9		5		2	8		3	
		3						
4								5

Easy 47

							8	
				9				
				3	4			7
	1				8		6	
5					6	7		3
	6		3		7		1	4
		5						6
	3	6						
4		8	7		1		9	

Easy 48

		4	2		7			
			4	9	5			
		3						
5			4	1		7		
		2			8			
	8			9				2
	1					9		
		9	7	8	3			
2						8		3

Easy 49

5	7		8			4	3	
4	3				5	8		1
1			4					5
	4					9	1	
		2	5		3			
	1			8				
		7					4	2
			3	7				

Easy 50

		5	9		3	8		
	2			1		3		
1	4				8		7	
							1	
4			2					5
	7		5	9		1		
2	8				1	7		
		4		8				2

Easy 51

				1	3	6		
		8						
7					4			1
	4			5			8	
		3		6				
						5		7
	3	2				8	7	
		7	5					4
6								

Easy 52

		5	2		3		6	
						3		8
							5	9
	8		5			7	3	
7		9			6		8	5
			9	4		5		2
				6	2			
					7		9	4

Easy 53

1		9	5			3	7	
3		5		1		9		
			6			8		
	9			6				
8	1				3			2
2		3	9	7				
			7	9		5	1	
					8			

Easy 54

			9					4
3		4			6	2	8	
9					8			
7	2							3
8	9		1				2	
					9			
6	3			8		7		
	7				3			
				6	4			

Easy 55

	3				8			7
8		9	4	7			5	1
			9					
		3	1		9	7		5
				4				
4			3	5				
			5					2
				2	7			8
7					4	3		

Easy 56

3		8			6			
6		1	4			8		
	9			8	3			5
2						3		
			9					
			2			9		
							8	6
	6			4		5		
				5	8	1	9	

Easy 57

					2		3	
					7	6		
	7				3		4	9
					6			
2			4	5				7
				8		3		
6							8	
	8			9				3
						2		4

Easy 58

			2	7				
		5			9			
			1	4	5		7	9
3		9	8					5
				3				4
4			5					7
				3				
		1	4	8				
9	3	8			2	4		

Easy 59

	8					2		
				4			8	
2						7	1	5
				7				
9			8		2	4	6	
1		2			9			
5			4			6	2	
		9		6			5	

Easy 60

			6	4				
					5			
	3				7			9
7		5	2	1	6			
6	2				4	7		
3						1		
		4	9	6			5	
5	9			7			3	
				2				

Easy 61

		8	7	4			1	
				8	7			5
						9		
	9			7				
2	1		4					
6					1	8		
					9	1	2	
		7						
			8				5	6

Easy 62

5		1		7	4			
	3	4		5		7		
6							2	4
			3					
		7						
2				1				
	6			8		3		
			5	3			4	
3			6		2			1

Easy 63

3	7				8			
		1						
	5		2	1	4		7	3
					2			5
	4	8	9	7	1			
7		4				9		
2	8				5			4
9				8				

Easy 64

			3					
		7	9			1	8	
				6	1			9
3	1							7
	4		8		3	5	9	1
				5				
	3							
	9	4						
5	7	1				3		

Easy 65

4	8							
	2	7	6			1		3
						5	4	8
	5				1	3	7	
6				7		8		
			2	5	4			
						6		
				2				
			5	1	3			7

Easy 66

					8			4
6	9					5	2	7
		2				6		
				5		7		8
8		3						
	2	5					3	
4	8			3				6
							4	
3		9						5

Easy 67

		7		1			5	
		9	2		8	6		
					5		7	3
6	7			9			2	
		2			1	5		
3						9		
						7	3	
	5			3				
	8		5					

Easy 68

5								
								4
2		7					3	6
	7			5		9	2	
		2		8			4	
6	8		4				5	
				7	2	3	9	
				8		9		5
				5				

Easy 69

		6		5			1	
1						7		
		3						
5		8		6		4		
	9	7					6	8
	3		5		8			4
	1				4	5	7	
	6			9		1		

Easy 70

	1							
		8		7	1			3
7						5		2
	5		1	2	7	8		
2						1		
			4					9
	2						8	4
9			7					
			9		3		5	

Easy 71

		6	1	2				4
1			7					
	7			6		8		
	4				2			
8	1	7	3					
5					6			7
							1	
	8		5	3			2	
2						7		5

Easy 72

8	7	4	5			1		
			8			5		4
		5		9				
		2		7				
9			3			7		
	3		9					
1	9		7	4		3		
	4			2				1

Easy 73

8	9	4				6		
		2						
	6							
		9			3	8		
2					9		4	
7			8	5		9		
			9	5	3			6
4			3			7		2

Easy 74

			1	8			2	
3					7		4	
	1	8			5			
			4			5		1
	4		6					
6	7		3					2
	6							4
							3	
	2						7	

Easy 75

		2		4			9	1
	9	7		6				
6				1				
8	4				6	9		7
		3		8				2
	1				7			8
			6			2	1	
						7		
		9						4

Easy 76

		5						
			3		7			4
	4			6		7		
2			9					
		6	2					9
		3	8				4	2
	9	7				5		
		4		2	8			6
6			7					

Easy 77

	7	9			8		6	
2		8		4			5	
		5				9		
	9	7	1				8	
				8			2	
			6		7			
	1				6			
		2		1		6	9	
	8		9	7				

Easy 78

			9			3		
6	1		5			4		
5				2		8		
	8				2	5		
9				6				
2	5			1				6
		4						9
			4	8	3			

Easy 79

		5	4					
3				8	2			
4	2			5				
		3				8		6
6		4						
			7		4		1	
			7	8	2	5		
		6						4
			1		5			

Easy 80

8				6			9	4
						3		6
	7							8
		1	9		3	2	7	
			8	1	7			
4				9		6		
		3						
7	6				2			3

Easy 81

			9					
			2	6				
6	4				3			2
	9					4		6
5	2		9	3			7	1
3	7			4	1			
		9		5				
	3		7	1		5		
2								

Easy 82

2			6	3	1	4		
		7	8					2
	8					6		
		8						
1		3			7			
	4							
	6			4				
			7	9	8		1	
8	9	2			6		3	

Easy 83

					3		2	
				7			3	
		4						
		2		4	7	8		
	9			3				4
		8			2		5	9
4					8			6
2	5							
7					6	9		

Easy 84

7	3					5		
	5		8				4	2
		2	6			9		
			5					6
	4	9				3	7	
				2				3
8	6		5					
	9			6	8			

Easy 85

	8			3	4			
		6	8			1	9	2
		1						
								3
	9			8			2	
			6	9		5		
	3					4		
			5				8	6
			2					5

Easy 86

5	3			2	6			
2				5				7
1		3			2	4		
			8					
	4	8		7	1			
7			6				2	4
		2		1	4	7		5
		1						

Easy 87

	3						1	7
		6	9					
	4		5		3			
		5		9		1		
		9		7		2	4	
	6		2					
							7	
					5			
	5	2			6			3

Easy 88

	5	9		2			7	
		1						
4					7	1		
		3		4		7		
		8		9			2	
8				7		3	9	
		7	4					5
		4	2		1			

Easy 89

		7	5		1	3	6	4
8					6			5
				7				
							7	
		1	9			6		
						4	8	
9	6							
			7	5			4	
3			1					

Easy 90

	4							
	2		4	6				9
								5
	3		1	5			4	
1			4			8		
	6			8				
	2							
				1		9	5	
	9	5	2			6		8

Easy 91

5		6	4					7
	1	3						
		4			8			
			2	7		3		
	7			1		2		
1	4					7		5
		5	6					
				3				
				5		1		6

Easy 92

			8					
	4		7	9		2		
		8	6	1			7	5
	6			2				7
7			5				9	
	1	4			6		2	
1								
								8
		5	2	4				

Easy 93

4								1
8				1	3	2		
5	3					8		
	1	2	4	3				8
	5		6			1		9
	4		1	5				
				2				
				9	4			
							6	

Easy 94

		5	3	6			8	2
				1	8			6
	7			5				1
						1	7	
		3			1	8		
7	5						3	
	2						6	
					2			
3			9	7				

Easy 95

						1	3	
8					6			
3	5					9		
			5	8				
		6	1	2				
5								
	3		7		8		2	
	7	2	6				1	9
			2	1		7		

Easy 96

	2	3						
			3			1	6	7
			7			4	2	
3		4			9	5		
	7	6					9	
9			6	5				
						2	5	1
				3				
			6				7	9

Easy 97

		5	3	4	9			
	3				8			
8						7		
7		9						
3	4		1					8
			4			9	5	
4	1					3		
5	9	2		7				

Easy 98

	1			5	9	6		
9								
	5	6		4		8		
				3				
				8		1	4	3
		1						
5			4					
6		8			3		9	7
	3		5			6	4	8

Easy 99

								4
6		3		4		2		
4		7		9		6	8	
			6	1	8		7	
							9	
						8	2	1
1			3					9
		4	9		1		3	
			4	8				

Easy 100

	5	1						8
9								
				4			1	5
7				1			6	
1		3		9		8	7	
	9		4	7	5			1
				8	9	3		
	7				3		9	

Easy 101

					3	8		7
1			6				2	
	4		2		8			
		9						
			3		1			2
			7	6		3		
6			1	9		7		8
						9	1	4
							6	

Easy 102

			8	7				4
			2					5
2		6			4			
9			7		1			
		4		9		2	6	
6		1						
	5		4					
		8						
			9	2		7		

Easy 103

		7	6	1				
				7				2
6			2		4			
2					7			
			5					1
		9		3				
		2	1					
4	5				2		3	
7	3			4	9		1	

Easy 104

			5	2		1	8	4
	5				8			6
			9			5		
			4					5
		3	5			2	6	
	6					3		
			8			6	2	9
3								8
2				4				

Easy 105

6		4				8		9
			9	8	1			
				4	3			
				4	5	7	9	3
			3			6	4	5
		9				1	5	6
			5		8			
3						9		8

Easy 106

7			1	3	4		9	
	2	5		9				
3				5		1		2
1			2			9		
	6						3	
				7				
			2	6		3		4
						6		

Easy 107

	5		7	2				
	2	7		9				1
9					6	2		
			4	6				
							1	9
							7	
	1	9	6					
4	3			1			5	
			2				4	

Easy 108

1				8		7		5
				7				
	8	9					2	3
2			4		8		9	
		4	1					
			7				3	
						9	8	
	4			5				
9			3				7	

Easy 109

3	7						9	6
1			8				5	
					9		2	7
9	5		2					
		6	5	1		7		
	3			8		2		
				9		6	7	
		5					8	
			3					

Easy 110

7						9		
			1	3				
	5							
		6	7	2			5	
	9	8				6	7	
	3						1	
	1	9			3		6	
			6					2
8				9		3		

Easy 111

	3	7				2		
9		8	3				5	
2	4						9	7
		3						
8						7	2	
	1	5						
		9	2	3				8
1				8	5			
			1		9	4		

Easy 112

5								
	2	4	7	5				
	6		8					
			9	5		1	8	
9								
					2	7		5
2		7	6	1				
8	1			4				
	9			8				

Easy 113

6	1			7				4
	7	3					9	1
		2	6			7		
9			3	6			2	
2		6		5			3	
				1			4	
			5					
				3				
	9				4			

Easy 114

1			9					2
								7
9		2	1	4		6	5	
	7			3	6		1	
					1			9
2								
			5			4		
				6	9		7	
6				2				

Easy 115

3					9			2
6		8				3		
							1	8
				4				
1	3						8	5
		5				9		
2		1		8				
						4		
9			5		1		6	

Easy 116

			9					7
						6		
		3	6	4			2	
						3	5	2
		6				1		
9	2		7			4		
	6	9		3				
		2		1	9			
7			2			5		

Easy 117

	5		4			8		6
8	6				5	9	3	
4								
			3			5		
		4		6	8	7		
			9					
	9		3	8			1	
6		1				7		

Easy 118

			8		6			
				1		8		3
			3		9		2	4
8						5	4	
6		4			3	1		9
		5		4				8
1			2		8		9	
						3	6	1

Easy 119

7					1	9		3
	1		6	2	7	4		
6								
	3							7
		1						
2			7	3				8
	4	9	8		3			
				9				
1			2		6			

Easy 120

8			6					
7	4						9	5
							1	8
	7			9	8		6	1
	1		5	2			7	
		9				2		
	9			5				4
	6	4			1	7		

Easy 121

6						5	8	7
		5						
			6			2	3	
5					7	3		
		7		6	2			9
			9		5		1	2
			7		3			
				1				
7		6			9			1

Easy 122

	3				7		1	
	4		6		2			
				3		4		
1								
		8		2			7	6
	7	2						
	8							
		6		7	3			
			1		5	2	4	

Easy 123

			1	6	7			
			3		9		1	
							2	
	9					7		6
		1		8	6			
	7						5	3
			6	9				8
5								
7			8		1			9

Easy 124

			3				5	4
		6	4			9	3	1
	4		1				8	
			9		2			6
1			5					
				5		3		
4	1		2					9
3						8		

Easy 125

					4	1	5	
		2			9			7
7	4		5			8		
	8							2
	7					5		
6		9				4		
		7	2				9	
5	1	8						
			4		1			

Easy 126

2								9
	5	1	9			6		
	6	9	7	4				1
5				1	9			6
	4		6		5		9	2
		7					5	
4						3		
								5
				2				

Easy 127

	1	7		6				3
6		2			3	1	4	
	3							
				4		2		
			9	8			1	
			6					8
3				9		8		
				3		6	7	9
					4		3	

Easy 128

			1		5			
7						1		9
	2	1		6		4		
	7			3	6			1
3				9				2
	4						6	
	5	9			3			
	1		2					
						3		

Easy 129

			6	4				3
		1						
	6				5		7	
2				5	1	4		
	8		4			2		6
	1			7			8	
4							2	
								5
		7					6	

Easy 130

		6	9	3				
		8		1	6		2	
						8		
			5			6		2
	6	3	2	1		7		
			7				9	
			5	9			7	
						5		
1						3		

Easy 131

7	6			1		8		
			9			6	4	
		9						
		3			4		8	9
			3		1			
	4		8		9			7
	3	6						
8	1					7	2	
						1		6

Easy 132

			8	5				
			3		4			
		6		9				
4				3			8	
		1			7			9
6		8					1	5
	1		9					
	5			6		9	3	
3				5				8

Easy 133

	3				4			5
					5	4	3	8
9	5			8				
	7				2			3
	8						7	6
		9	4					
5		6						
						2	9	7

Easy 134

	1			7	5			
	6							
				8		9	3	6
	8	1			7	3	9	
9		3			8	5		
6	5		3				1	2
			9	2			5	
5						6		

Easy 135

3	1	9	2	4	6		8	7
6	4		7	1				
9				3				
	3	4	1					
			6					4
4	7	3		9				
	6	1		7	2		9	3
				1				

Easy 136

			2				6	
				7	1			
5					8	7		3
7						8		
	9	5	3			2		
2		6	5	9			7	
				2		5		
	6							8

Easy 137

				4				
3		4	9			6		
	8					9		
		7	1	9				4
			3	2				
				6				
	1					9		
6	9		4	8		1	7	2
	7	2				4	8	

Easy 138

	4	9						
	1		3	6				
2	3		5		4			
6	9				7	5	2	
7	2		4					
							4	6
						1	9	7
1							5	
								4

Easy 139

5				7		6		
			2			8		7
8		7					2	1
		6	3		7	1		
		2						
		5	1		6	3		4
4					5			
		1	6					
					4			

Easy 140

			3					6
				8	2			
		2				3	9	7
	8	3		1	5			
								3
		5	6	7				
	1					5		
6						2	8	
			1	5	6			

Easy 141

	9	4				8	2	
			7	8			9	
7		2					6	
				7		1		8
		3		4				
				1				
4								2
		9		8	2		7	
3			4					

Easy 142

		6	5		1			
	1			7	6		5	9
8								6
1		9				4		
	3							5
6						3	9	
9	8			6			7	4
7						8	3	
				4				

Easy 143

			4		8	6		
	9				2		1	
		6	5			4	2	7
							6	9
			7	6	5	4		
						7		1
1		5						2
			2		4	9		

Easy 144

	2					8	5	
	1	3			9			
8					6	1		
4	3						8	
						2		4
		5						6
						9		
		1		9				5
6	8			2				

Easy 145

```
4 1 . | . 6 . | . 7 .
9 . . | . 1 . | . . 8
. . . | 7 . 9 | . 4 .
------+-------+------
. 7 . | 4 . . | 1 9 .
. . . | . . 1 | 5 . 6
. . . | 6 . . | . . .
------+-------+------
. . 1 | . 5 . | . . .
. . 9 | . . . | 7 3 .
. . . | . . . | 8 . 9
```

Easy 146

```
. . . | . . . | . 7 .
. 3 1 | . . . | . . 8
. 4 . | . . . | . . .
------+-------+------
3 8 . | . . . | . . 1
. 9 . | 3 5 . | 8 . .
. 5 . | . . 7 | . 9 4
------+-------+------
. . . | . 9 . | . 4 .
. . . | 4 . . | . . 7
. 2 4 | 8 . . | 5 . 9
```

Easy 147

```
1 6 . | . 2 9 | . 5 .
. . . | . . 2 | . . .
. 5 . | . . . | . 3 8
------+-------+------
. . . | 7 . 5 | . . .
. . . | 1 3 . | . . .
. 7 . | 6 9 8 | . . .
------+-------+------
9 . 6 | . . 3 | . 8 5
. . 1 | . . 6 | . 2 7
. . 5 | . . . | . . .
```

Easy 148

```
1 8 . | 6 . . | . . .
6 . . | 8 4 3 | . 9 .
. . 4 | . 1 . | . . 3
------+-------+------
9 . . | . 6 . | . . .
. . . | . . . | . 2 1
. . 6 | . . . | . . .
------+-------+------
. 1 . | . 5 . | 3 . .
. 4 . | . 9 . | . . 2
2 . . | . 5 . | . . .
```

Easy 149

```
. 6 . | 3 . . | . 1 .
. . . | 9 . . | . . .
8 . 9 | . . . | . . .
------+-------+------
3 . 2 | . . . | 9 5 .
. . . | . 5 . | 1 . 7
5 7 . | . . . | . . 8
------+-------+------
. . . | . . . | 5 2 .
. 8 . | 5 . . | . . 3
. . 6 | 7 . 2 | . 8 .
```

Easy 150

```
. . 8 | . . 9 | . . .
. . . | . . 5 | . 7 .
. . 7 | . . . | 1 . .
------+-------+------
. 5 1 | . 7 3 | . . 4
. 2 4 | 8 . . | . . .
8 . . | . . . | 5 . .
------+-------+------
. . . | . . . | 7 . 9
. . . | 4 9 . | 3 1 .
2 . . | 3 . 1 | . . 5
```

Intermediate 1

	2	9						
6	5						2	
			7	3				
			8			5	9	
1				2			3	8
		3	1		9	2	7	
8	3	2		6				
	6							
		5			8			

Intermediate 2

	1	9		6	3	2		7
	2					9		1
4							6	
						3	2	
			9					
								6
		1		4		6		
	6			3		4	5	2
7	3				5		1	

Intermediate 3

			4				6	1
2		8			1	3		7
			6			2		
7	3							
					3			
			1	4				8
4	9							
		7		6				
1			2		7			

Intermediate 4

			5					2
5		2						4
6			4	3		5		
	7			4				
9								1
	1					3		
	9	3					1	
						4		
4				1		6	7	

Intermediate 5

		5					7	3
4			6			1		9
2								6
		9					2	
		1	9			3		
3					5		4	7
5	9				6			
		2			5			
		4			7			

Intermediate 6

4			3		2			
6				4	5			
		4	7	8	6		5	3
8	5							7
								4
	8		5	6			9	
						2	8	
		9		3		7		

Intermediate 7

				3		7	2	
	1			5		6		9
2			9		6			
					2		7	
8	9						1	3
	3				7			
5			1				9	
		1			5			8

Intermediate 8

	1		6		5		8	7
8			4					
	4	7	9			5		6
6		9			8			
		1					3	
	5					9		
								4
5						3	1	8
				6				

Intermediate 9

		3		5		8	7	6
5	7					9		
4								1
	8		1		7	4		
								7
3	9			6	5			
							4	
			6		9			
						5	1	

Intermediate 10

				6		2	8	1
						6		
							4	7
6	7					1		
4		1	3					6
3								
	1		7	5				
	6		1	3		8	5	
	8	5		4				

Intermediate 11

			5			7		
		9			6			
3		4				9		
	7		2			3		
			7					
4			3					5
1		6	5	3			4	
9		3						
	4					2		

Intermediate 12

						5	2	
4	2							
6				4				
							4	
8			1	6				
							9	6
	5	9	8		6			
2			3	1		5	8	
					2			9

Intermediate 13

4								
	1	3						
	2			5			6	
			6		1	5	4	
9		4	2					7
				4				
			4		7		2	3
					3			9
2					6			

Intermediate 14

		1			9			3
	5	8				1		
								2
			6	3	8	7		
2			7				9	
		5		2				
3	9			6				
5		7						6
			8		3	9		

Intermediate 15

	8			5				1
7						4		
		5	8	7				
	3	9	1			6	5	
				3		8	7	
			5	9				4
	3		6					
6	4							
	7	1						

Intermediate 16

		5		8		2	6	
		3	7	6				5
		5		1				7
	3					1		
	1					6		
						2		
	8		6	4			3	
	2		8		5			
	5		1			4		8

Intermediate 17

	7							8
				5	6			
6	5		9					1
		3			7			9
			2			3		
		7	8					
			1		3	2	9	7
					9	5	1	
			5	2				

Intermediate 18

9	3					6		
				3				
7						4		2
			8					
6	8					7		
1							8	
3			4	7		2	1	
								8
	4			6	9			7

Intermediate 19

8				5				
		7	2		3			8
	1	5	8	7		3		
	7				5		3	
				4	2	1		
	9	1		2			5	
		4						
9		8						
	5					2		

Intermediate 20

	7					6	1	
3		8						9
	2		7		9			
6						2	4	
			4		8			6
		1						
	8					6	4	
2							9	
	6	9	8				1	7

Intermediate 21

								9
4			6				3	
5	3	1	4				8	6
	6			2			5	
	2				6			
			3			8		
6	5		8		4	9		
3	8		9					5

Intermediate 22

			6	7		8		
	5		8	3				
			7	5				
				1		6	4	
7		4				1		
			2			7	6	
6	8						5	
	4		1				2	

Intermediate 23

9		4	1	3				
6				5		7		
		3	7	6			1	9
7				9			4	8
			3		7	9		
						8		6
					5	4	7	
				8	6		9	

Intermediate 24

	1	4				3		
2	6						5	
		5	8			4	2	1
			9	2	6			
6							4	2
		3		6		8		
			1			5		
	4				8	2		
	8							

Intermediate 25

```
. . 4 | . . 7 | . . .
. . 5 | . 8 2 | 6 . .
. . . | . . . | . . .
------+-------+------
1 . . | 3 7 . | . . 5
. . . | 8 . 3 | . . 7
. 4 . | . . 8 | 2 . .
------+-------+------
. . . | 4 1 6 | 5 . .
. . . | . 3 1 | . . 6
. 8 . | . 7 . | . . .
```

Intermediate 26

```
6 7 8 | 2 . 4 | . . .
. 2 . | 8 . 1 | . . .
. . . | 3 . . | . . .
------+-------+------
. . 6 | . 9 . | 1 . .
. . 1 | . . . | 6 8 .
. . . | . . . | . . .
------+-------+------
. . . | . 4 . | 7 . .
1 . . | 9 . 6 | . . 4
. . 4 | . . 3 | 2 . 9
```

Intermediate 27

```
. . 4 | 7 3 6 | . . .
9 . . | . . . | . . .
. 2 5 | 1 9 . | . . 4
------+-------+------
. . . | 5 7 . | . . .
. . 6 | . 1 . | 5 . .
4 . . | . . . | . . .
------+-------+------
3 . . | . . . | 1 . 5
. . 1 | . . 4 | 7 9 .
. . . | . . . | 2 . 3
```

Intermediate 28

```
. 3 . | 9 5 . | . . .
. . . | . 7 . | 9 . .
. . . | . . . | . . 5
------+-------+------
2 4 8 | . . . | . . .
3 . . | 4 8 . | . . 1
. . 1 | 7 . . | 3 . .
------+-------+------
. . 5 | . 2 . | 3 . .
. . 2 | . . . | 4 . .
1 7 . | 5 . . | . . .
```

Intermediate 29

```
. 3 . | . . . | . . .
1 . 6 | . 2 . | . . .
. 4 2 | . 5 . | . . .
------+-------+------
7 . . | 1 . 2 | . . .
. . . | 3 . 9 | . . .
2 6 5 | . . 3 | . . .
------+-------+------
. . . | . 6 . | . . 7
. 7 . | . 4 5 | . . .
3 . . | . 1 . | 6 . .
```

Intermediate 30

```
. . 4 | . . . | . 7 .
. . . | . 6 4 | 8 . .
. 8 . | 7 . . | . 5 .
------+-------+------
. 4 . | . 1 7 | . 2 .
. 6 . | . . . | . . .
2 . 7 | . . 5 | 1 3 8
------+-------+------
. . . | . 2 . | 4 . 7
. . 3 | . . . | 2 6 .
5 . . | . . . | . . .
```

Intermediate 31

	9		1					
	4							
		6	8			2		
5			6					
			8	5		7		
8			2		7		4	6
		7				9	6	
				4	9			2
	2		5					

Intermediate 32

2		7			3	4		
	5	8			6			
6						2		3
			7	3			4	
						5		
		4	8				2	
		2			7			
		6	2			8	7	
			6	4	9			

Intermediate 33

		8				6		
		4		3				
3	1							
9		1	6			2		
4				8				5
	6					9	1	
			5					
2	3				9			8
	5		8					

Intermediate 34

		2			4			
						5		
		5					1	4
			3				8	
3			8	7	5		4	
9				1	2		7	
5	9						2	
	2				3	1		
7			1					

Intermediate 35

7	5					1		9
6		3		7				
	1							
	8	1		9		7		
	6			8				
						4	5	
	9	6	3					
1			6					
			5			9		7

Intermediate 36

	3					1		
							5	
			5	1				4
				6				2
9						3	4	6
	7			4				
4			7			9		
	5		3	2				
			9				6	7

Intermediate 37

7				8		4		
							6	7
		6						
			8	4				1
	5	1						
6		8						
4		7		1		8	9	
	1	5						
					5		1	2

Intermediate 38

			5	2				
	1	4						
			9			3	4	
3			7			2	6	4
5		1		3		9		
							5	2
					3			8
8	2			6		1		

Intermediate 39

		2	7			8		
	5		2			3		4
3			5	4		6		
			2		5	4		
8		1					2	5
6					8			
	6							
			1		8			
			7	4	3			

Intermediate 40

6						3		8
1		5	3				9	
	9	2			6			
2			6		4	1		9
					9			
		3			2	6		
					5	9	6	4
						5		
4								

Intermediate 41

				9		2		
		1						
		7	1	6		3		
3			2					7
9	7				1	5		4
		4	6					
6		3			5			1
1		2				4		

Intermediate 42

5			8		2			
7			9					
	3							
1		8			4	3		
3			7					8
							1	4
	2	4			9	8		7
	7				3	9	4	
								1

Intermediate 43

					4	2	1	6
						8		4
			5			9		
4	9			8	1			
						3		
			9		2			
		9		1				
1	5					6	2	
			4					8

Intermediate 44

		7	5		1			6
				9		5	2	
	2	6						
	6							
							5	
1	4				7			
2					4			1
	7		2	3		6		
	9			1				5

Intermediate 45

	2	5					6	
6					7	4		
						1	9	
						9		
7			2		4			
		6						1
			7	5		8		
5		9		8		7	4	
		8	4			6		

Intermediate 46

		8		9			3	
					6	4		
6			3				9	
7	3			2			4	
	9		6				1	7
			1					
3			9					1
				1	7			
			8				6	9

Intermediate 47

2			3					
		7	4			2		
	1	4	9			3	7	
					6			
					2	4	1	9
					5			
		9		5		7	3	2
							5	
	5			7	1		9	4

Intermediate 48

	9	8		7				2
7		1		8		6		
6	3		9		1			
				3	8			
3			1	2			6	
1			9	7				
				2	1			
		7						3
					4			8

Intermediate 49

```
. . . | 4 . . | 7 . .
. . . | . 6 . | 9 . 5
. . . | 2 3 6 | . . .
------+-------+------
2 . . | . 8 . | 3 . .
. 4 8 | . . . | . . .
. 9 . | 5 . . | 6 . .
------+-------+------
7 . . | 6 . . | . . .
4 5 9 | . . . | . . .
8 . 3 | 9 . . | 5 . .
```

Intermediate 50

```
. . 9 | . . . | . . 3
. . . | 6 . . | 8 . .
3 6 5 | 4 . . | . . .
------+-------+------
. . . | . . . | . . .
5 3 . | . . . | 4 . 1
. . 8 | . 1 . | 3 9 7
------+-------+------
6 5 . | 7 . . | . . .
. . 4 | . 3 . | . . .
9 8 7 | 6 4 . | . . .
```

Intermediate 51

```
. . . | . . . | 1 9 .
2 . . | . 4 . | . . .
. . 8 | . 3 . | . 2 .
------+-------+------
. 7 . | . . 9 | . . .
. 1 3 | . 2 6 | . 4 .
6 . 2 | . . 7 | . . 9
------+-------+------
. 2 . | . . 4 | . . 3
. . 6 | . . . | 8 1 2
. . . | 9 . . | . 7 .
```

Intermediate 52

```
. . . | . . 2 | . 8 .
. . 8 | 4 1 7 | . . .
. 3 . | . . . | . 7 .
------+-------+------
4 . . | . . . | . . .
. 7 9 | . . 6 | . . 3
. 8 6 | 1 . 3 | 7 9 .
------+-------+------
. . . | 6 . . | . . .
. 1 . | 7 9 8 | 2 . .
. 2 . | . 3 4 | . . .
```

Intermediate 53

```
3 6 . | 5 . . | 2 . .
. 4 5 | . . 3 | . 6 .
. . 8 | . . 6 | . . 9
------+-------+------
4 . . | . 7 . | . . .
. . . | . . . | 5 7 .
8 . . | . . . | . . .
------+-------+------
. . 7 | 6 . . | . 9 4
. . . | 3 . . | . . .
. . . | 9 . . | . . 3
```

Intermediate 54

```
. 1 5 | 6 4 . | . . .
2 . . | 5 9 . | . 3 .
7 . . | . 1 . | . . .
------+-------+------
. . 3 | . . . | . 9 .
. . 9 | . . . | . 1 .
. . . | . 2 . | . . .
------+-------+------
. 4 . | 9 . . | . 5 7
. . 2 | . 5 . | 4 . .
. 9 . | 4 . . | 1 . .
```

Intermediate 55

7	9	3				2		
		1						
4					9	3		1
8			6					2
3		6		7				
						3	4	
				6				7
6		7	3		4		1	
		2				8		

Intermediate 56

	5	3						
			8	5				3
8	2							
					3	4	2	9
6								
		8		4		5		
		5				9		2
	3	9			7	8		
4				3				5

Intermediate 57

	6	7	1	5		4	3	
					8	6		5
			2	3				
						1		
5	1							
		3		2		7	6	
	3	6						4
4			3					
				7				

Intermediate 58

					3	7	5	6
				4			8	
		8						3
	4							
	5		3				6	
	6	2			7		3	1
			1		6		4	
			5					
	3		8			1	2	7

Intermediate 59

		7			4		9	2
	8		3		5			
1				2				
9		4	5		2			
			8					
					9			1
						4	7	
			9			5	1	
3				7	8			

Intermediate 60

	6	9	3			2	4	
	7							9
4			2			5	3	
								5
	2			7				
	3		6					
								3
			5		6	9		
	9		1				6	4

Intermediate 61

				7		5		
1	9			2	5	3		
							9	
9						1		
		2					8	9
	5	7	8					6
5	8				3			
	7	3					1	
						7		

Intermediate 62

6			9					
	5		6	2			9	1
		8	7	5				
				7				3
	9	5	1					
						9		
	2			3				
1				6				
	7			2			3	5

Intermediate 63

								4
	7	6	4			8	5	
		1		8		7	9	
				5				
		5	7	9				
8			6			3		7
	6		1			4		
1		7	5				8	

Intermediate 64

			8	5		4		
				9			6	
			3					8
	5							7
		3	7	4		5		
	9	6			8			
	4			2			3	
			7			3	2	5
						9		

Intermediate 65

	5						4	6
					6		7	9
7	4		9			2		
9		5			7			8
8						6		
					2	8		
	2		8	4		1		
1				5				

Intermediate 66

6	1			3				
				8		9		
			7			4		1
				5				
				9				
3		9	6	8				
		8	3	6				9
	6		4		1			8
	5					1		

Intermediate 67

	6			9	5		4	3
		5	3	6				
		3	7		4		2	
						5		2
							7	1
5		2	1	7			9	
						7	6	
					1			
							5	4

Intermediate 68

5				3				
	7			5	1		9	3
9							2	
			7					8
		3		2		1		
			3	9				
						1		
7	1					8		
4		9		7	8	3	2	

Intermediate 69

5	7		9		8		6	
		3	7					2
9			2					
2		9				3		4
				2	5			
8	6		9	4		2		
3		4	8			7		
		8	5					

Intermediate 70

			3			2		6
4			7		5			
9			4					
	9	5			4	7		
2				7				
						6		1
	2	7	5	1				
				6			9	
						5		

Intermediate 71

6			8	7				
4	7		6				3	
9		3		5				
	9			2		8		
7	4		9					5
		6					7	
8	5				4			7
3		7	2					

Intermediate 72

	6	3		9	2			
4		7	1			6		
8		1			7			
7	4					9	2	
		8						3
			6	8		1		
			3	6		7		
			9			1		
						9		

Intermediate 73

	8							
	4	5				9		
9		1		6		5		7
	6	9						
	2		5	7				6
8								
1			9			8		5
		6	1				7	2
			7	4		6		

Intermediate 74

	1		8	2				
		7			4			
5			1				3	8
	6			5	7			
							5	2
			6			3		1
			6	4			1	
							2	
		5			8		4	6

Intermediate 75

						8	1	
	8			6	2	5		4
	3					2		
8			5	1				3
			6					
	1			3				
		5					8	
7	6				4			
1					3		2	

Intermediate 76

5	1	9				2		6
	8						1	
		6				9		
	4		9			5		3
	3							
1					3	8		
			4		8	6		5
6	5		1					4
8								9

Intermediate 77

1	3					9		
			9	6				4
					4	2		
								7
		5	4			6		
	6							5
8			5			4		9
6								
		4	3	6		1	5	

Intermediate 78

		2		7		8		
3				4				
	1				2	7		
			9	6		4		3
	3					2	9	
		1						8
				6				
	7	6		3			2	
		8			1			

Intermediate 79

4		2			6		3	
		8					5	4
5						1		
		6	9					5
2				1	3		9	8
	4	9						
		1			8			
	8				4			

Intermediate 80

		2						9
	4		8		9			
			4	6				
						5		
	9	4			8		3	
3	8		5					
			2	3				6
		5						3
9		7				2		

Intermediate 81

		9	1			2	7	
	4		7			9	8	
2					8	4		
			8	4			1	
		3		9				
			3	1			2	
		2			7			3
9			6			7		

Intermediate 82

	8							
1	6			8				
			3	5			1	
			7					
					1		9	2
	3	2	8					
	7		6	2				9
5		1			3	6	2	
6								

Intermediate 83

	3			2	4	9		1
		2	5		7			
5						4		
						7		
	2	7		1				9
					5			3
6	5			3	9	7		4
				2				5
						1		

Intermediate 84

	1	5	3					
	4			2	1		7	8
		7	9					3
		1	4	7		3		
							4	
				4	5	2		
	8			3		7		4
	5					1		

Intermediate 85

			3			5	9	
6					7	8		1
7		5		8		6	4	
						1		
				6		4		
							5	8
	7							
3			1	4				
4		8	6		5	9		

Intermediate 86

				4	8			
	1				5			
	7	3	9					
		1		5	4	7	8	
				2		9	4	
5		4		8		1	3	
7	5							
		8			2			
						8	9	

Intermediate 87

6			8	5	2			3
3				6				
	8	5						
							1	
			5					
	9	4	1	3		6		
		8	3			9		
4	6			2	9			
	2					1		

Intermediate 88

		6	2			4		9
		3				1		
	9			4		6		7
	7							1
	2			6		8		
			7				9	3
	3	1	9		8			
6								

Intermediate 89

	1	3	2	6				
7								9
		9	3					
9	3		5					4
	7			2	6			
						9		
		2	6			5	1	3
			4	3			6	
						4		

Intermediate 90

		1						
2				5				
			8					
3			6			5	1	8
						2	9	
		2			7	6		
			7					5
	8	3	1					
	5	6		8	2	9		

Intermediate 91

		2	5	6			3	
				8				
3						2		
	8				4			
		1	2					3
9					8			
	9			4	6			
			1				9	4
		6	8				1	

Intermediate 92

8				9	6			
								5
9		6				1	7	
	3							
5		9						4
4		7	3		8		5	
	8			1		4		
1			7					
3	9		4					

Intermediate 93

8	1					5		
		2	6					1
	7				5		8	
7	5		1			8		
				7			3	6
						2		
			2	3	8		6	
4			7	6	1			

Intermediate 94

			9			3		
	8				1	6		
								2
		8						5
1				2		9		
	5						2	8
			9			1		
2		6	7			3		
9				5	6			

Intermediate 95

7	8		5					3
			9	3			5	
					6			
8	9	1	7		4			
					8			6
	7		4	8				
3	5	9			7			
	1		6			9		

Intermediate 96

						7		
		8				9		5
			3			6		
		6	9	1				
		3	8			4		
	4		5	6			9	
					9	1	5	8
6	7					3		
5			4					

Intermediate 97

		8				6	9	
5	9						3	
			8	5			1	
7	8			6				
	3	1						7
			9	4				
					5		4	8
					1	5		
3		6						

Intermediate 98

	9	2			6			4
1	6			3				
							1	2
	4	1	6		5			
8	5				2		6	
		9		4		5		
					9	1	4	3
					3		2	
						9		8

Intermediate 99

					8	5		
		1	9		2			
3	2	5		7				
						3		7
						6		
		7	1					9
	8		7			1		
			5					8
	1	9	8			2	3	

Intermediate 100

		3		6		8		5
	6			2	1			
	7	4	5					1
				1	4			
			3					6
7	4		6					3
	3			4	8		2	
	1		7	5				
				3				

Intermediate 101

2							8	
			2					
	3	4		1				6
3		9				6	1	
	8	6						5
1					3			
4		1			6			2
			5			9		
	9	2		3			4	

Intermediate 102

	4	6	5			3	8	
				7				
2	7		3	8		9		
				4			3	
	6					2		
			8	3				
			7		2	5		8
						6		9
			9			2		

Intermediate 103

		1	4	7			2	
7				9	6			
					1			
	4					5		
	7				2			
8	5		6					1
						2	4	
9			7				5	
	8		5		9		1	

Intermediate 104

1					2			6
		7	1	8			5	
2								
		6					9	7
			7	5				
6	1				3			9
		8		5			7	
	3		6					1

Intermediate 105

	5		4		2	1		7
					7			2
		3		1				5
8		2						9
3								
			8			3		
		8				9		3
7	2		9		5		8	
			2			7		

Intermediate 106

2			3	6				1
						2		
3	9	4					8	
		1	9				7	
7					4		6	
	8					3	2	
		9	8					
			4	3	7			
		6		2				

Intermediate 107

		8		2				9
				4				
2	3	5				7		
5				3				6
	9		7					
6	8		9					2
		2			6			
	4				9	8		
	5		6					3

Intermediate 108

			2			7		
					6			
	7	6	4			9	5	
						3	9	
	5			9	8			
	3	8	7			2		
	6							
7				5		4	3	
	9	2				5		

Intermediate 109

		1	8					
				3	9			8
				9		7		
	5		3			1	4	
9		6	1	4				
	8					3		
			6					
	4	5						1
	8	9				4		5

Intermediate 110

			2				6	
	3							
			4	7				
					4			
3	1		8		6	4		
2	8		1				3	
				1	2			
1			3	8				6
4		9	6			8		

Intermediate 111

	2	9						
	5	6		4	9			2
			1	9			7	6
3								1
		7		5	4			
	6	2		4		1		9
				2			6	
	7			1	3			

Intermediate 112

			5	2			7	
	9	3		4		1	5	
7								9
	4	2				6		
	1							
		6						
2			3	9				
			6	1	7			5
			5		7			

Intermediate 113

				8		2	3	
			4		9			
8		7						
			2	9	1		5	
4					1			
	2							
	1				3			
7		5					8	4
3		2			7			

Intermediate 114

					8	7		
			1	6		5	2	4
	9						1	
		9				4		8
		6	4				5	
		2		9			6	
			2	7		6		
		5						
		5	9		8			2

Intermediate 115

```
. 7 6 | . . . | . . .
8 . . | 3 . 6 | . . .
. 1 . | . 5 . | . 7 2
------+-------+------
. . . | 5 . 2 | 7 8 .
. . . | . 4 . | . 5 .
. . . | . . 8 | . . 4
------+-------+------
. . . | 1 . . | 5 . 3
. 6 . | . . . | . . .
. . 3 | 7 . . | 8 4 1
```

Intermediate 116

```
. 3 8 | . . . | . 5 .
. . 9 | . . . | . 3 .
5 7 . | . . . | 4 . 6
------+-------+------
. . 5 | 3 7 . | 6 1 .
. . . | 9 . 6 | 5 . .
. 9 . | . . . | . . .
------+-------+------
. . . | . 6 . | 9 4 7
. . . | . . 8 | . . .
. 4 3 | . . . | . 6 8
```

Intermediate 117

```
. . 4 | . . . | 2 . .
9 . . | 2 . . | . . 7
. . . | . . . | . 3 .
------+-------+------
. 8 . | . . 9 | 4 2 .
. 7 . | . 5 2 | . . .
3 . 9 | . . . | 5 . .
------+-------+------
. . . | . . 5 | . . 3
. . . | . 4 1 | . . 5
. 5 . | 7 . . | 1 . .
```

Intermediate 118

```
. . . | 4 . 5 | . . .
. . 6 | . . 7 | . . 9
1 . . | . . . | . . .
------+-------+------
. . 2 | . . . | 3 . .
6 4 . | . . . | 2 . .
. . . | . 6 . | . . .
------+-------+------
. . 9 | 7 2 . | . 3 5
. 6 . | 3 . . | 4 . .
. 1 . | . 9 . | . . .
```

Intermediate 119

```
. . . | . 1 . | 8 . .
5 . . | 7 2 . | . . 6
2 . 7 | . . 6 | 1 3 .
------+-------+------
. . . | 6 . . | . . .
3 . 1 | . . . | . . .
. . . | 1 3 7 | . . .
------+-------+------
. . . | . . . | 6 . .
6 2 . | 9 8 . | . . .
. . 3 | . 7 5 | . 2 .
```

Intermediate 120

```
6 . . | . 8 . | . 2 .
. . . | . . . | 9 . 4
. . . | 1 . 9 | 6 5 .
------+-------+------
. . 7 | 9 6 . | . . .
. 6 9 | . . . | 4 . .
5 . 2 | . . 4 | 8 . .
------+-------+------
. 2 . | . . 8 | . . .
. . . | 5 2 . | . . .
. . . | . . . | 5 1 .
```

Intermediate 121

9				1		5		4
		8		7			3	
			4					
	3	1						
					4			
	7	5			3			
	1		9	8		6	5	
	9							3
5			6			9		8

Intermediate 122

			6	7			1	2
				2		3	4	
6			1					
	1	2					6	
		5		1				3
3		6				5	7	
		3	7			8	5	
			3		4			7

Intermediate 123

	1			9				
	7		8			4		
6		8						
	6		5			3		
	4			1	6			
8			9			1		7
1	3	5						
					7			
		4			9	6		

Intermediate 124

						1	6	7
1				4		8		
		7	2					
8		6		3	7	5		
	1					4		
7								2
5		1		7			4	
6		3	8		2			

Intermediate 125

		8						4
3		1		2			9	
		2			7	1		
			2	3				
					1			
					6			
2	4	3		7				9
8			4		6			7
		6				8		

Intermediate 126

	5		7	3				
	3		8					4
						6		9
		6			3			
			4	5		2		
		2		8				7
		3			9	5		
			4				9	
			2					

Intermediate 127

```
5 . . | . . . | 2 7 .
9 7 . | 3 . . | . . .
. . . | 1 . . | . . .
------+-------+------
. . . | 2 . . | . . .
. 6 4 | . 7 . | . . .
7 5 . | . . . | 9 . .
------+-------+------
2 . . | 6 . . | . . 1
. . 7 | . . 1 | 3 4 .
. . 6 | . 9 . | . . .
```

Intermediate 128

```
. 8 6 | . 7 5 | . . .
2 . 7 | . . . | . . .
. . . | 9 . . | 8 2 .
------+-------+------
. . . | 6 . . | 3 . 2
. . . | . . . | . 1 .
. 5 3 | . . . | 7 . 6
------+-------+------
3 . . | . . . | 2 . .
. 1 8 | . . . | 5 6 .
7 . . | . . 8 | . 9 3
```

Intermediate 129

```
. 2 . | . . 3 | 8 1 .
. 1 7 | . 2 . | . 9 3
3 . . | . 8 6 | . . .
------+-------+------
2 . . | . . . | . . .
. . . | . . . | . . .
. . 6 | . . 7 | 1 8 5
------+-------+------
. . . | 7 . 9 | . . .
. . 1 | 8 . . | 3 . .
. . . | . 6 . | 2 . .
```

Intermediate 130

```
. 2 . | . . 4 | . 7 5
. . . | . 6 . | . . 4
. . . | 8 7 . | . . .
------+-------+------
8 . . | . 9 . | 5 . .
. . 2 | . 3 . | . . .
. . 6 | 5 8 . | 9 . 3
------+-------+------
. . . | . . 3 | . . 6
. . . | . . . | 7 8 .
. . . | . . . | 2 . 9
```

Intermediate 131

```
. . . | . . 4 | 2 . .
. . . | 8 . . | . . 5
. . . | . 5 . | . 7 9
------+-------+------
9 . . | . . 1 | 7 . .
. . . | 9 . . | . . .
. . . | . 2 . | 4 . 3
------+-------+------
5 7 . | . 1 2 | 9 8 .
. . . | . . . | . 5 .
. . 2 | . 9 8 | 3 . .
```

Intermediate 132

```
. 9 6 | . . . | . . 5
. . . | 5 8 . | 1 . .
. . . | 6 9 . | . . .
------+-------+------
. . . | . 3 . | . . .
. . . | . . . | 5 . .
. . 5 | 9 . 7 | . . .
------+-------+------
5 . 8 | . . . | 7 1 .
1 7 2 | . . . | . 9 6
3 . . | . 8 . | 2 . .
```

Intermediate 133

			8		6			
2				4			8	
9		1					3	
			7			8		
	9	7				2		
	1		2		8			4
	3				1			
	2	6	7	9				3

Intermediate 134

					5			3
3			2					7
	6	5				8		
		8	7	2	6			
	2							
			4		8			
7			5	6	3	1		
1				4	5		2	
		3						8

Intermediate 135

			8					
	7		2			8	4	
		9		6			2	
			3				8	7
9	8				2			4
			1			6		2
		3	7					
	6	4			3		9	
				4			6	

Intermediate 136

	2	9				1		4
			4	7			5	
	8							
	4		7	8		2		
				1				5
2	1			4		3		9
3	9					5	2	
						9		
							3	7

Intermediate 137

	4							6
9	7		3	2			5	
8	2		9		4			3
	5							
	9	8						
		4	8	3				
5				7		4		
					6		2	
	8		2		9			

Intermediate 138

		3	9				7	5
		5			4			6
	8							
9	3			1				
	5	8			6			7
			3					4
			8					9
		6	7	9			4	
				4				3

Intermediate 139

								4
					1		7	8
3		5						
2	7				8		3	
						8		1
			5				2	
5		3	1		4	7		
					9			
1		4	2					9

Intermediate 140

4				8				9
	8		4	2				1
		1			3			
		3	9					
9	4							
				7			2	
3	6							4
7		2				8	6	
	1							

Intermediate 141

3	4	2				1		
6	8			2				
					7			3
			1				7	
				3	6			
		6	8		9	3		4
		1		6		9		
		8	2	1		6		
					4			

Intermediate 142

		4		6				
	9		1		8			
		6				5	8	
8		5			2	4		
			8				7	2
				4	1		6	
			2			6		1
		7					9	
9								

Intermediate 143

9	2					6	7	
				8		2	4	9
		4						
6	5	4						8
				3			6	
7	9						3	
2	3			4			5	
			2			7		6

Intermediate 144

				9		7	3	
3				2	7			8
8				1		2		
9								
5	3		8					
	7							3
				6		1	9	
	1							2
7	6		5					

Intermediate 145

		5		6		3	7	
	7		1				9	
9			4					
3				1				
6						4		3
	5				2			7
								1
	3				9	2		
		9	5					

Intermediate 146

			2				3	
				5		8		
1			6					7
	8	3						
	7		1	6	5			
			7					3
				7				
		8		5			1	
			8	3	4	6	2	

Intermediate 147

9					6			
	6					4	2	
8		7		2				5
		1		9	5			
2						9		
		6	7		1		5	
1			2			6		4
7				4	9	2		

Intermediate 148

5							8	
	9							
			9	6	1		4	5
			1			4		
	3							9
6	8	9			2		1	
	5	8		2	6			
		1				6		
9			4					

Intermediate 149

	3						1	2
		2	9				6	
				3		9		
	1							
8				1		2		7
6		7				8	5	
	9	6		8			2	
	7					6		5
	8		1					

Intermediate 150

	5		4	3		6		
	9							7
6			9	2		5		
3	6			9			2	
	7	5		6				
	4	1				7		
							3	
	3		6		2			
4						1	7	

Expert 1

			9				5	6
7		1		8	6			
			3			1		
				1		8		7
		3						5
		9				4	1	
6	5				4	7		1
	4						8	
1								

Expert 2

		9			4			
	7							
	5	2				4		
2								4
	6				2		8	
		7	9				1	
	4		2		8		5	9
	9	5		1			6	2
7			6			8		

Expert 3

	3		8	5		1		
	5						8	4
9								
								7
4		7	9	1				5
				4		9		
		8	5					1
5	6		7	3	4			
	4							6

Expert 4

5			4	1				
		4	6		9			
8				7				1
	2		9					
		1		8				
		8				9		
			2	5		1	6	
	5		7			2		4
2	9						5	7

Expert 5

						7		
		6		5				
3		9			2			
	3							
	2				1	6		9
9		4	6				1	7
		2				3	7	
				1	6	9		5
				7	3	4		2

Expert 6

		7			6		4	
								1
	8			5		7		3
						9	6	
	9	6			1			4
								8
4			1		8			
		8		6	9		5	
7			4			8		

Expert 7

		9	8	5		7		
				3		1		
	5	6	1		7	9		
	8	1	7				4	
	6					8		9
			3					1
7	9					5		3
						4		
8			9					

Expert 8

		8	3			5		
								7
			4					
		5	2	6			3	
8		3				7		
		7	8			2		5
7			5	3				4
	4				8	9		6
						3		

Expert 9

	4	8		1				
	1						8	6
		9			4	5		2
			8	2				
8	2	1		7	9			
	6					7		
5							9	
			2	5	6			

Expert 10

	1	2						
		7	2	6		5	3	
								9
3		8		5			9	
	9				7		6	
7				3	2		8	
		9			5		7	3
1	6							
				8				

Expert 11

	3		5	7			8	9
				4		3		
			2			1	7	
2	5		7		4			
3	4		8					
9		8					1	3
	2							
8	9		3	5				
								5

Expert 12

				2				1
	8		3		4	7		6
7			6			9	3	
	6							
	2	8	7					
	7			2	1	6		8
8		3			7		4	
			8				9	

Expert 13

4			8	2	6			
			4			1		9
2								
1		6		5				7
		5	6	7		8		
	2							
5	4							
8		9	2	1			5	
							7	

Expert 14

1		8			2			7
		4		1		3		
3	7				5			6
			6			7		
		2						
			1	4		8		
							6	5
	3		5	6				
8			2			4		1

Expert 15

		2						6
	8					5	2	7
	3		1		7	6	8	
		1	6		5			2
	5	3		6			1	
	1	8	7	3				
		6		1		9		3

Expert 16

		3		7				
4			6		8	5		
2			5	1		3		7
			2					
3		4	1			2		
8						1	7	5
	7					4	8	
				6				1

Expert 17

						6		9
					5		4	8
	2					3		
6		9		1	4			
		5	9			2		
8				9	3			
	9	1		4	8	2		
	4		5					

Expert 18

		5				1	9	2
				1	8			6
				7	6	5		
						2		
	5				7		8	
3								1
9				1			7	
7		6	3	9				
1								3

Expert 19

4	7		5	8			6	
					6	1	5	
2					1	7		8
7		3						
	1	4						
	2		7			4		
		2				6		
5			4		9			
	6					5		

Expert 20

							3	6
	8	2			1		5	
3			5	6		4		
8	6		4				7	
5								4
		7	8	5				3
				7				
						6		
7				6		8		

Expert 21

		6				4		
	3							8
	4		6				5	1
			4	1	6			
		3	2	8				7
6			5			8	2	
			1		4			3
5								
				5		7	1	6

Expert 22

						4		3
			8				7	
		5					6	
		8	5		2	6		9
				3		2		4
2		9	7	6				
		7	3			5	9	
	4				8			6

Expert 23

			2	6				
7	9	8		4			2	
			9	7				3
				8		3		
9						1	7	6
2			7		6			
		2						
4								
			8	3		2		7

Expert 24

4					2			9
						8		
1			2	6		5		
5	6		8	9				1
9		8		4				6
8	3					5		
			3					
2	9			1			6	3

Expert 25

	6				7	5		
			6	3		9		8
		9			5	3		7
	1			7	3			5
	7		1		4			
	3	8				1		
					6			
		5	4				3	
						4		9

Expert 26

			3					8
						9		4
					5		1	2
	8			1		6		5
9						8		
		2	4					9
	1			2				
	3			5				
6	4		9		8			

Expert 27

6			8			9		
		7				4		
2			6					
	1			9		6		7
					1			
	8		2				9	
1			4					
8	6					3		
	7	9		6	5			2

Expert 28

			8			3	1	
			7	1				9
				3				
	1		2				3	
2			5	3	6	9		
							7	
3		2						
	6	1		9				2
9		8	6	7				

Expert 29

		1			9			
		8		4	1		9	
9			3					2
7			2					8
	1	5			4			
					5	4		
	9				7	5		
		7	8					
	3					2		

Expert 30

						5	8	
	2							
9	1			4				
	6	1			8	9		5
			2					4
	4			5	3	2		
6	3							
			3					8
		5	1					9

Expert 31

5		2	4			1		3
	6			5				
			3	1			5	
				4				7
4	7			2	1	6		
8		5		7		2		
6								
						8	4	6
		8						1

Expert 32

			4					
						2		8
				5		7		
8				4				
1			6			5		
			7		8			3
	3		5			6		1
2		4		1	6			
		6	3			4	2	

Expert 33

			3			5		
		8			6	1		
2		1	5					
	8	7						5
9			6			7		
	3				7	6		
		9					1	3
6	2			1	3			9

Expert 34

		9	3		5			8
				8				
			4					
	3					9		5
		4	9		7			
			2			3		
		3			6		4	
2	8						6	
	5				3	2		7

Expert 35

5	7					9	6	
			6	8				2
						4		
			8		5			
	4	1				5		
7				2			4	
		2				8	9	
						1		4
6	1	9						

Expert 36

		2	1					
				2		4	1	6
			8	5		2	7	
8	4		5		6		9	
		6		1	8			
5								
	8			5				
7				4				
	6	5	7	2	1		8	

Expert 37

	3						9	
4								2
7		6	4	9		5		3
9			6					
	6				7		4	
			8	3				7
	2		3		8			
		8						
	5			7			6	

Expert 38

8	3				9	1	5	
5			7		8			2
9	4				3		8	7
								9
3	8						2	
		9	3					
			5		1			
	5						1	
	2	3	8		7		9	

Expert 39

6	1			5			7	
			1		9	5		
4			7	8				
	7	1				4	5	
8							9	
	9					6		
								4
				8				
7		8	2	6		1		

Expert 40

	8		4			2	1	
			1				7	9
		7		3		8		
8				4			9	1
9								3
	4	3			7			
		9				6		
			3					
2				8				

Expert 41

	4		8					
7		2						4
3	5							
			5		7	8		2
				9				
					2	3	7	
						2	8	
9	3			2		7	4	
		6						5

Expert 42

1			9		8	3		
		7						
4				5				8
		5	3			7		
			4				1	
							4	6
		6					9	5
				1		6	3	
7						1		

Expert 43

```
. . 4 | . . 5 | . . .
. 1 . | . . . | 6 . .
7 . . | . 8 4 | 3 . .
------+-------+------
4 8 2 | . . . | 1 . 5
. . 1 | . . . | . . .
6 7 . | . 1 . | . 4 .
------+-------+------
. 3 . | 5 . . | . . .
. . . | 8 4 . | . . 1
2 . 8 | . 3 . | . . 6
```

Expert 44

```
9 . . | . . . | . . .
. 3 . | 8 . . | . . .
2 7 . | . 4 . | . . 5
------+-------+------
. . . | 5 . . | 7 . .
6 . . | . . . | 9 . 8
. . . | 2 . . | . 6 .
------+-------+------
. 5 8 | 4 . 9 | . 2 3
. . . | 3 7 . | . . .
. . 9 | 5 . 2 | 6 . .
```

Expert 45

```
7 . . | . 1 . | . . 4
. . 2 | . . 8 | . . .
9 . . | 2 . . | . 3 .
------+-------+------
2 . 7 | . 5 9 | 4 . .
. 8 5 | . . 7 | 9 . .
4 . . | . . . | . 5 2
------+-------+------
3 . . | . . . | . . 9
. 5 . | . . 1 | . . .
. . . | . 4 . | . . 8
```

Expert 46

```
. . 5 | . . . | . 2 .
4 9 7 | . . . | . . .
. . . | 6 . . | 3 . .
------+-------+------
2 . . | 8 . . | 6 . 7
7 6 . | . . . | . . 8
. . . | . 4 . | . . .
------+-------+------
. 4 . | . 6 . | 9 . .
. 7 2 | . 4 . | . . .
. . . | 9 . . | . . 5
```

Expert 47

```
. . . | . . . | . 2 .
. 9 . | . . . | 7 . .
3 . . | 7 . . | 1 . .
------+-------+------
. 3 . | . . 8 | . . .
2 . . | . 6 . | . . .
1 7 8 | 2 . 9 | 6 . .
------+-------+------
. . . | 4 9 2 | . 1 .
. . 6 | 3 . . | . . .
. . . | . . 1 | 8 9 .
```

Expert 48

```
3 . . | . . . | 5 . .
. 9 . | . 4 . | . 6 .
1 . 6 | 2 5 . | . 9 7
------+-------+------
4 . . | 9 . . | . . .
. 6 . | . 2 7 | . . .
. 7 4 | . . . | . 3 5
------+-------+------
. 3 . | . 6 . | 9 . .
2 . . | 1 . . | . 4 .
```

Expert 49

					4			9
9	5	6						
		3				2		
1			3		2			4
	2		1			7	6	
	6	9	4		7		5	
	9			1			7	6
							2	
				9				

Expert 50

	5				2	4		
7				6				1
			8				9	
2			6	1		9	5	
			9	2	8			
					4			
		7						
	1		5			8		2
9			1		6	7		

Expert 51

9	3	6			1			
2					4			9
						1		
			9	5				3
5			4		2		9	
					2			7
4			3					
		9				7		2
3		1				6		4

Expert 52

	6	3					4	
5						2		
2	4	9		5		1	8	
	1			3				
							9	
9		6				4		
		8			1			
			4	2		9		8
1			3			5	2	

Expert 53

		9		2		8		
			4			3	9	
	4			8	7			2
		7	5					
			8					
	9			3				
							3	7
		4	7	6				5
5		2			4	6		8

Expert 54

6							1	
					1			9
				9				
9				5		4	7	
7			4	1				2
		6			9			8
					5	8		1
2			9			7		
	8			6			5	

Expert 55

```
.  6  . | .  .  . | .  5  .
7  2  8 | .  .  . | 1  .  .
.  .  5 | .  .  . | .  .  2
--------+---------+--------
.  .  . | .  7  . | .  1  .
8  4  . | .  .  . | .  2  7
.  .  . | 8  1  . | 3  6  .
--------+---------+--------
2  3  7 | 6  .  . | .  .  .
.  .  6 | 5  .  . | .  7  3
.  .  . | .  .  . | .  .  8
```

Expert 56

```
.  .  3 | .  .  7 | 4  .  8
.  .  . | 8  .  . | .  .  5
.  .  7 | 1  2  . | 6  .  .
--------+---------+--------
6  .  1 | 7  8  2 | .  .  .
8  5  2 | .  6  . | .  .  .
.  .  . | .  .  . | .  .  .
--------+---------+--------
7  .  6 | .  1  . | .  .  4
1  .  . | .  6  . | .  .  .
.  4  . | 5  .  . | .  .  .
```

Expert 57

```
.  .  7 | 2  .  . | .  .  .
.  .  . | .  .  . | 5  1  .
.  1  . | .  3  6 | .  .  .
--------+---------+--------
1  9  . | .  .  . | .  .  .
.  2  . | 5  .  . | 9  .  .
7  .  . | .  .  . | 8  .  .
--------+---------+--------
.  .  5 | 8  .  . | 6  2  .
.  6  . | .  2  . | .  3  .
8  .  . | .  9  . | 5  7  .
```

Expert 58

```
.  .  . | .  4  6 | .  .  .
.  .  . | .  .  . | .  .  9
.  .  . | 2  .  3 | 4  .  .
--------+---------+--------
6  1  3 | .  .  . | .  .  .
.  .  . | 1  5  . | .  .  .
.  .  5 | .  4  . | 8  .  .
--------+---------+--------
.  5  9 | .  .  . | 3  6  .
1  .  8 | .  .  . | 5  .  .
2  4  . | .  9  . | .  .  8
```

Expert 59

```
.  .  . | 6  .  . | 4  .  5
.  .  5 | .  3  . | 2  8  .
2  6  . | 4  .  . | .  3  .
--------+---------+--------
8  .  . | .  .  . | .  .  .
6  .  . | .  .  5 | .  .  .
.  4  . | .  .  . | 3  .  9
--------+---------+--------
.  .  . | .  2  . | .  .  .
.  .  . | 3  .  7 | .  9  2
.  .  . | 8  9  6 | .  .  4
```

Expert 60

```
.  .  . | .  .  . | .  9  .
6  .  . | .  .  9 | 1  5  2
.  4  . | .  .  . | .  .  7
--------+---------+--------
.  6  . | .  .  . | .  .  4
7  .  . | .  5  . | .  .  .
.  9  . | 2  .  7 | .  6  3
--------+---------+--------
.  .  5 | .  7  1 | 6  .  .
3  .  . | 9  .  2 | .  .  5
.  .  . | .  4  . | .  .  .
```

Expert 61

5								
						6	7	
		6	7		2		1	
			3	1			6	2
			8			7	3	
			5					8
6		3					2	
	1			8				
4	7				1	8		

Expert 62

							4	
6		5	7					1
9		4	1					
						2		
							5	4
			2		6	8		
	2	7	9					
				4			1	9
	8				1		2	

Expert 63

	2	6						
								1
	4	5	1				2	6
	5		2			1		4
8					5			
	1		8	4		3		5
4								
		9	6	5			1	8
	8	1		9				

Expert 64

	8	3		6	9	1		2
			1	2	5			9
					8			
			6		1			
						6		1
		9	2					
	4							
2		8			3	9	4	
6							3	8

Expert 65

					7			1
	8	4						
			6					
	7				9			
	5		1		8	6		
3				5				
			5				4	
9		8			4		3	5
			3			1		7

Expert 66

9	7							
1		3	7		6			
		2		3				
	1	8						4
				1		3		
			8		9			
	3			8	1		9	
				2	7			
			9				2	6

Expert 67

	5	2					8	
6			3	2	5			
4						9		
5				6				
	8	4						
					9			
	2	3			4			6
			6		3	1		
			2	8		5		

Expert 68

	1							8
					4		7	6
		6					9	1
8		3	4	1				
		7	3					
	5			8	9			
			5					4
6			7		8			
3		1		6				

Expert 69

				6		3		
1	7			3		9		
7					6			
		3	5				2	1
						4	5	
		6		2	7			
					9		3	
	5			1	3		7	2

Expert 70

			8					
8	1				2	5		
5				1				
			2		7			8
	7							
					4		5	6
			1				2	5
4							3	
	6	2		3			1	7

Expert 71

3			4	7				
	9		1	5		7		3
				3		8	5	
	4				6			9
9		5	3					
		6						
	3	7	9					5
			7			3	4	8
	1							

Expert 72

	6	8		5			7	
9	7		8		6	2	4	
	2		1					6
		2	6		5	8		
				7		1		
			2					
		6						4
	1		5					
	5		7				6	8

Expert 73

		4			3		7	
6			9		5		4	
5			2	7		6	9	
		7					3	
1	5							
4		3						2
2	6			4	9			
	1		7					
						3	2	

Expert 74

	3	1	9			4		
	2	7		5				
		6		2			7	1
			7	1	4			
							9	
		2				9		
	9				1	6		4
	6			7				2

Expert 75

5	8	1				2		
	2	7	1			5		
	6							
						7		
7		3		6		8	1	
8	3				6	4	7	
6				8	3			5
			2	7				

Expert 76

	9	2	7		4	6		
3		1		2				9
6	8			3	9	1		
					1	8		2
	7		6	8				
			1	9			8	
						2		
				4		7	1	6

Expert 77

	6	7				1		
8	4	1					2	7
			4			6		8
					1	2		5
7				9			1	
5	2					7		
	8		9					
		6	5				4	

Expert 78

			7			5		
			1		2	3		9
				3			6	2
6		5		7				
	9		2				1	
	2	1			6			3
						6	2	
8				1				
						9		5

Expert 79

5	8		1					
	9	4	7	3				
2			4	8				
		5		4		1	8	
			5					7
3								4
		7		2		5		
	5						7	
		8				3		9

Expert 80

		4	1					6
3							8	2
2				3	9			
		5	8					
8	9			6				
	5		6	9			4	
				2			1	
	1		5			8		

Expert 81

								6
9						2		
	3		7				5	
		3			9			
			6			7	2	8
1								
6	5				7		3	
			1		6			9
	8		2	3		5		

Expert 82

4			3					
	2						7	3
			6			8	2	
			4				6	2
9				3		5		
	7						8	
3			9		4			
			2			7	5	
	6		8					

Expert 83

		5				4		8
4	8			1				
						2		
8	4	7				6		
2	9						1	
	5			8	6		9	2
	6	9				1		
			1	4		5	6	

Expert 84

	8	3					9	
						6	2	
2			7					
7		5			3	2		
	2			9				1
3				8	6	9		
			6	7		3	1	
								8
					9			

Expert 85

		1	6	9			3	
	4					9		
				3		2	6	
	1							6
6			2				7	9
2				4		8		
	9		8					2
					1			
			9		4	3		

Expert 86

9							1	
	4		2		6			
				5	4		8	
			1	7	5	4		
7			4	2	8			
		2		4		9		5
	8				1	2	7	
6			7			8		

Expert 87

			3			2		
6		4	1					
		3		8		7	6	
	5	7				8		
1				3	5	4		
		1	2		7			3
		8		6			1	

Expert 88

		3						
9	5			7		6	8	
2			8			5		
3						2		
5			3	8	2			7
					9			
	2	4			8			
7		6				4		
	3		4				7	

Expert 89

7				3				
3			2		6			
								2
2	8	7		9		1		
	9		1		4			
						9		
4	3			2				
6		9		4	8	2		
	2				9		3	7

Expert 90

		3	7	5				1
6		1			3			4
		5		4	6	9	3	
				7		8		
					8	3	5	
	5	4					1	6
	6		3	8				
3				6		4		

Expert 91

4			6	3		1		
			8	7				
		7			5			3
		1	2	8			3	7
	3					8	5	4
						2		1
						5		
			3					8
8			1	5		6	2	

Expert 92

		9	3	4		5		
		7		5		8		
	3		9			8		
5	4					2	1	7
9		3						5
8	5						7	
							4	
3				1				9

Expert 93

			4			8		
1						4	6	
								9
8				9		1		
				6		3		
3		4		1	8	7		
7	3		8					
	9		3	7		6	5	
			9					

Expert 94

	5		8			3		
7		9						
			9		4			
	8					5	4	
				9		2		
		4		8	3	1		
	9	1	3					
	2						5	
				1	7			

Expert 95

5								
	2		9			5		
	7	9			2			
				4	8		2	
			7					5
8			2	3			9	
4				6				8
2	6				5			3
	8		3			6		

Expert 96

7	8		6		1			4
9	4							1
5						2	6	
		8	1		2			9
			7	4		5		6
4				9				2
	1			2				
			4		8			
		4						8

Expert 97

	2					1		
					4			6
					7		2	8
		2	7					5
8	1		6			9		4
		4		5		8	6	
			1		6			
							8	
	4	7			2			1

Expert 98

	1		2				9	
		5				2		8
				9				
								9
			1	6			4	
		4		2				
		8		6			1	5
	3		9		1		6	
	2		5	8				

Expert 99

	3	6						5
	5		9			6		
				3				6
		1		8	7	5		
2				7	9			8
7			1			6		
			3				1	
	1	9	8		5			

Expert 100

1			8	6				
			4					9
		6		2		5		
				5	4	8		
				9			4	
9						7		6
			2					8
5				7		6		
		8			6			4

Expert 101

	7			8			1	3
		2						
3		1	5					
			2		9	3	5	
		8	4		7			2
	9					7		
						5	2	
				3			8	1
		7	1			9		

Expert 102

					4	3		
3	4			7				5
7		2	9			4	8	
						9		2
5			3	2		7		8
2			8	6	7		4	
				5		6		3

Expert 103

		6			2			4
5			4			9		
	4		6				3	
				6				3
		4			9		7	
						5		
2		7					6	
9				3			5	
				2				8

Expert 104

	5			4	6		2	3
					8	5	7	
4			2				8	
								6
9			7					
3	6			5				
2					4			
8	3	7						
5						8	6	

Expert 105

				2				
7		5	1					
						4	5	7
			8				1	
	8					3	2	6
3	1							5
	3		7		8			
1		4						3
	5						4	

Expert 106

6				5	2			
				6				
	5		1		8			
								3
	1		2	6		9		
2	6	3		8				1
	2				5			9
4			8	1				
	8					5		

Expert 107

			1				7	
1	7		9		2			
				7				9
	9				8			
5				6				
			7					6
		5			1	8	9	
9			3			7		2
6	8						1	5

Expert 108

			2		4			
	4		6	5			7	8
	3	5		8		2		
		4	8					9
6				9			3	
					7		6	
	7							8
		8						
		9	3			7		

Expert 109

```
. 3 . | . 8 7 | . . .
. . . | . . . | 9 1 6
. . 9 | . . . | . . .
------+-------+------
5 . . | 3 . . | 6 . .
2 6 . | . . . | . 5 .
. 8 . | . . 1 | . . .
------+-------+------
. . . | 2 8 7 | . . .
. . . | 1 . . | . 8 5
. . . | 7 . . | 3 6 .
```

Expert 110

```
. 1 . | . 2 . | . 7 .
. . . | . . 8 | . . .
9 . 5 | . . 1 | . . .
------+-------+------
. . 4 | 9 . 3 | . . .
7 . . | . . 5 | . 9 .
5 . 3 | 2 . . | . . 1
------+-------+------
2 . . | 8 5 9 | . . .
3 . . | . . . | 5 . .
. . 9 | 3 . . | . 2 7
```

Expert 111

```
. . 7 | 5 8 . | . . .
9 . . | . . . | 4 7 .
. 4 3 | . . 6 | . 5 2
------+-------+------
. . . | . . 9 | . . .
2 . . | . . . | . . .
. 7 . | 2 6 . | . 4 .
------+-------+------
. . . | 3 . . | 5 . .
. . . | . 9 7 | 8 3 4
. . . | . 4 . | . . 9
```

Expert 112

```
5 . 2 | . 6 1 | 3 . .
9 . . | . 3 . | 2 . 5
. . 3 | . . . | . . .
------+-------+------
. . . | 8 . 7 | . . 6
. 8 1 | . . . | . 5 7
. . 5 | . . . | . . .
------+-------+------
1 . . | . . . | . . .
. . . | . 6 . | . . 3
. 5 8 | 7 . . | 1 . .
```

Expert 113

```
. . 9 | 6 1 . | . . .
6 7 . | . 5 . | . . 3
. 4 . | . . . | 2 . .
------+-------+------
. 1 . | . . . | . . 4
. 6 . | . 9 . | 1 . .
. . . | 2 . 9 | . . .
------+-------+------
. . . | . 3 . | . . .
3 . . | . . . | 7 6 .
. . . | 6 7 . | 2 . .
```

Expert 114

```
9 . . | . 6 . | . 7 5
7 . . | 9 4 8 | . . 2
6 2 . | . . . | 8 . .
------+-------+------
. . . | . 8 . | . . .
. . . | . 5 . | . . .
. . 7 | . . . | . . 6
------+-------+------
. 5 . | . . 7 | 4 2 9
. . . | 4 . . | . 8 .
. . 1 | 8 9 . | . . 7
```

Expert 115

					3			2
	3		4					
9	8	4						
		9		1	8	3		5
			3				2	
3	1		2	5		4		
			9		1			4
	5	8						
1					6			

Expert 116

	8	7					2	4
6			3	7			5	
2						6	1	
	1							
		2	6	4				
8			1	5				
	3			1		4		
				3			7	
			5			2		

Expert 117

	9	1		2	7			
			1			4		
	7		6	3	9			
7					2	3		5
9			3					
2		4					9	6
			5					
				7	4		6	
					6	7		4

Expert 118

3	4		6					8
		1	3			6	4	
2		6		1			9	
	8	2			4		1	9
		4		8				7
	1			9				
	3				7			
1								
				6		4		

Expert 119

	7		9	2				
8								
			7	6		3		
	2	7	6	5	8			
9	5							3
3		6				7		
			3	7		6		
6	9			8				
		1		2				

Expert 120

				3		4		9
	1	3						
				5				8
2			4		7			
				9	8			
							9	5
	9					3	5	7
	8				4			
7					1			2

Expert 121

			1				3	
					4		6	
6		4			3	2		
		8			5	3		
		3	2			4		
4		6		3		8	5	
2	3				8		4	
8				6	9			

Expert 122

4			6				9	
		9	2				3	
	6	4			3			7
		7	1			5		6
9							1	
			5	2	4	7		
		2			7			
	5			1				

Expert 123

3	6	8			4	1		
		9						3
			1	9				
2			9			5		
4								6
1		6		4	5	3		8
	4			3				
			1		2		4	

Expert 124

8		5		3		1		
			1				5	
							9	
9		1	3	8				
		8	5		2		3	
				9				1
	4		9	5		6		
5			6			3	8	2
						9		

Expert 125

	9						1	
4			7		5	6		
5		6				3		
	4		8		7	9		
	3	1	5					
9		8		6				
			9		1		5	
3			4					
		4		8				

Expert 126

7				8				
						6		
					2		4	
	3					7	6	
				4				9
8						3		
6		3		7		5		
5	4		2		3			
	7	9			6	8		

Expert 127

				8				4
	7	4				2		3
	3	8			2	1		5
				6	7			
	5		8		1			
		7		5	3		1	
		5						
			1			7		
			6	3	8		5	2

Expert 128

	9							
6			1		7		4	
			8	4	6			
		3				2		
9			4	3	8	7		
	7		2			4		
3	4	2	7					9
								1
				6				

Expert 129

	6			8			9	2
				2		4		
			6				8	
					4		5	1
5	2							
		8				6		7
	9		5				2	
			1			5		
2				6		1		9

Expert 130

		2					6	3
		6		9			2	
		9	3					5
	6	7	8			1		
	3	5		7				
			2			8		
			9		3		5	1
2	5			6				

Expert 131

2	4			7	6			
1	6	7						
		6		2			8	
			4		3	2		
		3		6		4		
			9		2	3		
8				3			9	
	9	2			4	8	6	

Expert 132

			8				3	
5			7			8		9
		3						
			3	4			2	
			9		6	7		
				5				
	4		2			6		
9	6					3	7	4
	7				9	5	8	

Expert 133

3			2		6			7
2	9		5			8		
7			1	8				
			3				9	
	6					2		
	2					3		8
				5		6		
				1			5	2
					9	7	3	

Expert 134

	7			3		5		
1	6			4	5	7		2
				2			1	6
	3	7			4			
2				6				5
6	1						4	
	2	4	7					8
3				2				

Expert 135

5	1						7	
				5	8			4
			7				3	2
1								
					9	7	4	8
8			3	4			9	
		8			3		2	9
	5		4		1			3

Expert 136

	4		2			3		5
9	6			5	8			
		8			4			
2				4		7		8
				7				
						5	9	4
7		2					3	9
4	8							
3						2		

Expert 137

	4	3		5	9	7		
7						8		
9		5	7	1		4		
			9		4			
				8		5	3	
							9	8
	7							
							4	1
			4	3	6			

Expert 138

9	2						3	4
			3			2	8	
	4				5			9
		3		9				
	5			7	2			1
		4				9		
			2					7
				7				
	9	5			3		1	

Expert 139

2	9			5				
		7	2	9	4	1		
					3			
3	6		4			9		
		9		6			5	
1		5				3		
5	3						6	
				7	2			

Expert 140

		2	9		5		7	
		6						
	7	5	8	1			4	
8							5	
	1		7		8	4		6
		9		2				8
	9	4	6			2		
2			4			5		

Expert 141

6							4	
4		9	2	8		7	1	
7								
					9	1		
	7					9	2	
					3		8	4
	2		7	3		8	6	
1		8					3	
						4		

Expert 142

6			7	5		2		
7	5	3	1					
			4	6				
			2	8				7
			5		7			
4		1						2
	7		3			1		5
3					6	8		

Expert 143

	6							
			3			2		
1				5		4		
	3					9		
7	1		9		4			5
	4	2		7		1	6	
3			6				7	
					5	6		4
			7		1			

Expert 144

			5	4				9
	7		2			6		5
				7				1
5		4			9			
2			4		8			
4		1	9					
	5		1			8		
		9	7			1		

Expert 145

	6		5		3		8	
	4				7	1		2
			4					
			6		4			
2		8		5		3		4
	5	4	1					7
	7	2						
3		5						
			2					

Expert 146

1	4		8		6	3		7
	8		7					
	9		3	4				
			6				3	
	5	8		7		9		
7			9				6	1
						1		
9								3
						4	9	6

Expert 147

				6	1		9	2
	9	7						8
			9					4
6	7			8				3
								6
	8		4					
	6			3	2			9
1						6		
3				4				

Expert 148

							1	
			8			6		9
			3	6				
				3				4
9	5				4		8	
3							9	
8		9	6	5		4		
		1		7				8
7		4	9				6	

Expert 149

		2	6					4
	8						1	
		9		1	2	3	7	8
1	3			6				
2								
						9		
				2	8	4		3
		3	9					
6				3	7			

Expert 150

				4			2	
						5		
9	3							
					2	6	9	
		1		6				
	2	4					3	
7		2		6				
			1	7		2		
3	6		9				1	5

Expert 151

				9				
		1	8					
7	5			4		1		
		6			4			
	7	9		5		6		1
2	8							5
			9	2		7		
6							9	
			7		6	2		

Expert 152

			5	1		2		
6			9					3
		3						
3		4	9					
	5		2			1		4
			7	4				
7			4					2
			6					
	2					5	9	

Expert 153

		1	7			9		
		2			4		3	7
8					9			
			4					
6					2	1	8	
				6				
						8		
		8		7	1			4
		4	3		8		2	1

Expert 154

			5	6		9	3	
						7	2	5
8								6
				3		6		
5			7			8		
			8					9
	8	5	2	9				
1	2		8					
	6			1				

Expert 155

1		6						
			1	5		9		
3						6		
7	2		3	1		5		
			9			7	2	
		7			3	1		
5	9					2		3
		3				8		

Expert 156

						8		
	1	4				5		
				9				
	2	9	8	4	3			
5	4							
		6						
			3					2
	8		5			9		
			1	8		6	4	

Expert 157

5	9	4						
								1
		8			2			
			2	5	9			3
			3			9	6	
	8	3			6			
	1		9					
		2		4	1			
	3				5	8		4

Expert 158

5						6	3	
		9						
			4					1
		4	6				5	
	8						4	
			3					
1				6				3
	5					8		4
	3	6	1			9	7	

Expert 159

	9					8		
	2	5	4			6		
			7		3			
								5
		4			5	2		7
7						9		
			3	6				9
		8	4					2
4				7		5		

Expert 160

	2					7		
9					8			1
		8	4	5		3		2
			3	5		1	9	
3		9		4				
						2		
1					3	5		
	3		5					
8								7

Expert 161

2			6	1		9		
			7					
	6		4	9			7	
						2	8	
9				5			1	
6			2				5	
			1					
	4		8	6		5		2
				2	9			7

Expert 162

			6					
3				5	4	9		
		4			9	6	2	
2							9	
	5	1						
	6						7	
					1			9
		2			7	4	1	
			3		2		5	7

Expert 163

					7			
3		2						
6			8	2	5	3	7	
	8	4						
5		3			4	1		
2				5			8	
4		6		1	8	7		
1								
	3	7		6		2		

Expert 164

		9						
	3	6	9	7		2		
	2					7		
8								
			2				4	1
1	7		3	9		6		
	1							
				2			3	9
					8		1	

Expert 165

7		1			3			
		4						
3	9				2	6	7	1
8		2	7	3				
				4	1			
1			2		9		4	
	6		9		8	7		
								4
			1	6				

Expert 166

	9				8	1	7	
4				7			2	8
			2					
9								6
2						4	8	3
		6		4			1	
6		1	9			8		
	7					3		

Expert 167

	4	7						5
	3				5	1		
6	1	4						
9					7	3	4	1
				9	3			6
7		5	4	6				
			5		1	8		7

Expert 168

			4			2		1
6		8		3				
3	1			7	5			
	6					7	8	4
	3	5			8			
			6	1	7			
			3			8	7	
					4	5	1	
								6

Expert 169

	4			2				8
6				4				
2				8		4		
1						6		9
			9	1		2		
			7					
		1	8			4	9	
	3	7						
	8		3					2

Expert 170

4			7					
				1	9			
			6				2	1
	4		9			6	7	
	8	6	5					2
		2	1					4
			7			6		
2		9				1		
			9	4	5			

Expert 171

					7			9
	2		6	9		3		
	5		3	1				
				2		1		
			8	3	6			
	9		5					8
	1	5				9		
	7		1				8	3
	3	9			6	2		

Expert 172

5			6				3	7
1				8				4
						2	1	
3	5			7				
			6					
	6		2	1	5			
	7			5				
		2			4	5		
			8		1			

Expert 173

	2	7		8				1
8								
4		1	3				5	2
							1	4
								5
	1		5		7	6	8	
				8				
7			6				2	
		4	1		3			8

Expert 174

3				4			6	
4					8			1
		2	6			8		3
6	2							
	5							9
					9		2	
		6	4	9	1	2		
9		1	8	2				5
			5					

Expert 175

	4		9	2				5
8					1	2		
		4		1	9			8
	5							
	6	2	8	5				1
	8				3			
	2	1	6					
				9			3	

Expert 176

	2		1		8			4
1			9		6	8		3
6	9				2			
			8					
		4	2				9	
				4	5			
			5					
9		6			1			
2		1	6			5		

Expert 177

	4	2	7		3			
		3		4	2			
5								
	2	9	3					
				5			7	
3	6				8			
		6		9		8		
2						4		5
			8			9		

Expert 178

8			6	5				
					1	7		
		9		8				
4		6			7		5	
	3		8		6			
	1			3				
		4	3				1	
				7		4		
7	5			9	8		3	

Expert 179

5			9	8				
			1				9	
	1		6		5	2		
8				1				5
			4			6	1	7
				9	6	4		
					4			2
			5	2		8	4	
		6						

Expert 180

					1	9		5
	3		7					
			1	3				
		2					8	
8	5			3	6			7
9		3		5				2
						8		1
	6				2			
2		5						

Expert 181

					5	8		3
			6					
	6	4	1			7		
		3				2		
	2		4					5
				3		6	8	
		5	2					
8					6			1
	3			7			5	

Expert 182

3								
	9					7		2
1		2					5	
				6			9	
					1	8		
	2				8	1		7
		6					1	5
7	1						3	
		5		8	2			6

Expert 183

9							2	
	5	3		2	8		7	
7	8					6		9
		7	6		2			
3			7	5				
				3				
5				1				
	1	6	5		9			2

Expert 184

3						4		
		8	4	9				
1								
						9	4	
			3		9	8	2	
2			6			5	1	
	3		2		4		8	
5		4		6		3		
	2		9		3	1		

Expert 185

		3	7					6
	4	9						
						8		
8		5				6	1	
	9				6		5	
1	3		8		7		4	
7	1			6	3			
					4			3
		8						

Expert 186

					1	9		
4		7			2	8		
			9	4				7
				5				
		1				7		
3		2			8			
		9	4					
	3			8		2		
2	1						5	4

Expert 187

5		3		6				
						2		
7	9				5	4		
9					6			
	2	5			9			
3		4	2				9	
2			3					
8	4		7			9	2	
					4			

Expert 188

6		8						
3			6	5	2			
2				4	8	5		
	3			6				5
				7				
		6	3	4	2			8
4			5	8		2		
1						4		

Expert 189

		5			2			
7								
		4	7					5
			7	1	2			
		7	1					9
5			9		4			
	1	9		3	2			
			4	9	6			
						1	3	

Expert 190

5			7	1	6			
		4				1	3	
9			3					
		7		9				8
3			4			9		
8				6				
1						4	5	
7	8	9	1					
4		5		8				

Expert 191

	5	2		8		1		
		8	6	1	9			
			2		4		8	
					7	6	1	
4	7			6				9
		1				2		
							7	
		6				5		
	1	5						

Expert 192

7	6		5			3		
3	2	4		7	1	6		
	5					2	1	
6						5	4	
	1							
5			1	3				6
	4			2	7	5		
		9	6	5		2		

Expert 193

				2			4	5
		8	4		3			2
		3		7				9
	3	7	5	6				8
					8			
	5		3		7			
			2	8				
9					4			
	7	5			6	8		

Expert 194

	5				9		7	2
					3		5	
7	8				2			
						2		
6				9	7	8		3
			8			9	1	
							8	5
			3			6		
8	1			7				

Expert 195

			8				7	6
8		3	5	7				
					5			
9		7				8		2
				3				
2	5		8					
1						7	6	
5	8		1			2		
				6				1

Expert 196

9	3	1	2				7	6
			9			2	3	
6							8	
				1	5			
8	1		7					
7		2						
1		7	3		6			8
				8				
2			1	7				

Expert 197

						8	5	
		6		5		9		
9		3	7					4
6					7		9	
					5		3	
	7			1	6		8	
	4	9				7		
				9		5		1

Expert 198

	9		7			6	3	
8				9				
4				1			2	
				6	7	4		
		8		3	2			
				8				
	3				6			
	8	6	3				7	9
		4			9			2

Expert 199

		6				3	7	
3				9		4		
8	4	3		5	1			
				7	4			
4	8	1						
	5	6				9		
							6	
	6	1	9	4		8		

Expert 200

		8			6			
			5			4		
2		7			8			
		6	8			7		
	4	9			7			3
		5				4	2	
						3		5
6			3					
	8			9				7

Expert 201

	4					1		
1	7	5	8					
		8				2		
5		7					9	2
6			2			5		
		9	5					
	1					2		
				4				6
		2	7		5	8	1	

Expert 202

		4			3	9		
			5	4				
	5			7	9		4	
						1	9	
	2		9		7			
1				5	6	3		
			6		2		1	7
			7	9		2		
				3		4	6	

Expert 203

		6	7			5		
3					9		2	
9			3		2	1	8	
						3		
6		1						
2	3			7		6	9	
1		5			6	8		
	6	3						
7			9					

Expert 204

7		9						5
			5			3		1
1		7	8			5		4
	2			4				
5			1	7			8	2
			3					
8					2		7	
		2	5				1	8

Expert 205

6				4				
	1						3	
3			1	8		2		
			9	7				
1		2		6		7		8
							2	6
						3		
4								7
8						6	9	4

Expert 206

			5			4		2
			3		1	5	7	
		5		7			6	8
5		8				2	3	
7	4							
		6	1					
			6		2			
2			8		5		4	
				4				

Expert 207

							4	2
4	2					3	7	1
	3			9			6	
			4			6		
7	6			5	3		2	
	4		1	7			5	3
			7					
5								
				2				5

Expert 208

	7		3					
					4			
5	6		7					
		4	1	5		7	2	
		5	6					3
			7	2				
8							4	
	5	1				6		
		6	5				3	2

Expert 209

						2		
	6		9				3	
2		9			6	4		1
					1			
1		2			8			
			7					
	4			3		8	2	
	3	7		4			9	
				6			7	

Expert 210

5	1				7			
	7	6				9		2
			2		6			9
				3				
	2	4			1			
		3		4	5	6	2	1
		1				5		4
			9			7	3	

Expert 211

4				5	9		1	
			2	3		8		6
				6				5
						2		1
5		2			4			
			9					
					5			4
	5		8					3
6		3				9		

Expert 212

			5			1		
3	9				2		4	
	2			7	4			
				7	6			
6			3				1	5
			9	4				
	3	2			9			
	7			6		4		
4								

Expert 213

8		3						5
6			8	4				9
		2						4
			7	2	9			
			4		3			
				6				
	2	4	6					3
9		7		3	8		2	
		6		5		9		7

Expert 214

	1		9		3			
		5						9
9	7				2	6		
			8			5	2	
							1	7
				6				8
			3	7			5	
7				8				2
5			2			1		

Expert 215

		5			9			
2			6	7		4		
			1				6	
			1	4				
			2		3	5		
4		6	7	9		3		2
3			9			7		4
1					2			9

Expert 216

			8		6	2		
								4
1	3			7			8	
	2		7					5
				5	2	7	4	1
	8						1	7
			3		5	4	2	
3				2			5	

Expert 217

			9					
	6	1						
4		5	1					
6	8				5	9	2	
			3			1		
		2						
2	3		6	9	1	5		
	1		5		8			4
5				3				

Expert 218

		4	1				5	
	8	1			4			
5		7						
	2							4
			9					8
	1					7	9	
8			2	1		6		
	9		5					2
			8			9	4	

Expert 219

		9					4	7
5					1		3	
		4					9	
8							5	1
			6					
			8	7				
	1		9	5	8	4		
			7		4			6
		5			6			3

Expert 220

			8	7				
			6				9	8
			1	9	5			
7	8		9	4				1
	5	2				7		4
			9	2				
		1	8			4		7
6						2		

Expert 221

	5		1					
7				9				
	1			2				
1				6				
6			3			5	7	
							3	9
	7		6				9	
4		6				2		
3			5		7	4	6	

Expert 222

		7	9		3		2	
6		2		7				3
	4					1		
4								2
7	8							
9		6						
			3		7	8		6
	6		1				9	
			8		9		1	

Expert 223

2		6				8	5	
9	4		8		5	3		
	8							6
6	9	1		2			8	
	2		5	1			4	
4								
				4			2	3
			2		1			
							6	

Expert 224

8		7	9		3			5
		9		7		4		
2	3							
						2	3	
			5					9
7	9				2			4
3				2	5		4	
					6			7
6		4	3				8	

Expert 225

9		1				4	3	7
							9	
			6					1
1							4	
		9	4				7	
6				8				
		8		5	3			
3	1		7	6				8
					9	7		

Expert 226

1		7				4		
2	4		1				3	8
				8				9
3		2	6					
	9				2	1		
	1		8					
			3					
				9		6		
6		1	2	4			8	

Expert 227

7	2		3		5			
1	4			8				
			4	1		7	9	2
			8					
			1			3	5	
				7		4		
	3				8		1	9
2								3
8	1						7	

Expert 228

				5				2
2			6	3	9	5	8	
	3			4				
6					5			9
		7					2	
		9		7	4			
				9	2	4		6
	8						7	

Expert 229

1	6	5		8		3		
			9	5		7		
9			3			5	7	8
		8		9				
	3	6						
	7							6
					9	1		
3		4				8		5

Expert 230

			4	3				5
		7		9			8	4
6			8				7	
5		6						
	7							6
3	9	8				7		
				8		4		
7				4		1		3
		3				8	5	

Expert 231

6		8			4	2		9
			6				7	8
	4							
	1							
		4	5	9				2
8	2		1	7				6
		5	2			8		4
			5			1		

Expert 232

1				4		9		
		6	1			3		
4		3	6					
						5	2	4
2	7			9				
	6		3					7
			7					9
7	5							2
			6					

Expert 233

			3			5		
				1	2		3	7
			5	6				9
8		2	7	3				5
6						7		
						6		8
2		7	8			9		
				1		8	2	
				9		5	3	

Expert 234

	2	5	6					
				7				
			2	8			7	6
	7		3	4				2
		8	6			1		
5		3	1		7			
								3
4	3			1				
		6						

Expert 235

	6	7	9	4				3
4						2	1	
	8		3			7		
					4			
			7	1				
						8		
1		2	8	6				7
	4				2		9	
	7	8		3				2

Expert 236

				6	5			
	7	9			4			
			9			2		
	9		6					5
		2	5	7		3		
			3					
6	8			4		7		
5			8	7		4		9

Expert 237

					4		7	5
		1	3				4	
	6			7			1	
3			5					
7					6	4	2	
				6		8		7
4					8			2
	3	6			1			

Expert 238

	2		8	5				
		5		3				
4					9	8		
5							3	
		3		2	6			
				4		9		
	5				3		8	7
8		4		7				
7				9				6

Expert 239

			2		3			8
					4	3		
1			9	8				
5							1	
		7			2			
	9		8					
4					1		3	
	2			9			4	
	3	8	4			7		

Expert 240

		5			6		4	
6								3
				9				
5	8			9				
7		4		3	5			1
1				8				
						8		7
		7					9	
	7			5	3	1	6	4

Expert 241

		6		1	2	7	4	
7	1		3			8		6
						1		
			7	3				
			4	2		5		3
			8					
		8	5		7			1
	6	4						
1					6	4		8

Expert 242

	9	6		3		2		
		3	6					4
5	7	4				1		
		2						9
		7	2		9	6		3
	5							
	6		3					
	2	9						5
					1	7		2

Expert 243

	5	3						
			1	4				
4		8		7				
7							4	
				5				
	6	5				1	7	8
		4		8	6			
	7		1	5				3
				2				5

Expert 244

			7					1
	6		4		2		7	5
3	5							
	3			6		2		
5	8					1		7
4			1			6		
			5					6
			6			3	2	
				3	4			

Expert 245

			1		7			
6		9				5		
7							1	
		2		6		9		
				8				
			9	1		3		
	7				5			1
8	3	1				5		
5			3			7	2	

Expert 246

	1	9			4	6		
								3
	6							1
	9	3	1			2		6
		2	9		3		1	8
5				4	2	3		
		4		2		9		
					9		5	

Expert 247

			5		6	3	9	
					7			
			9					
3			7	1	5		8	
	7	5			2	1		6
9	8						3	7
2	3					8		
		6						9

Expert 248

					7			
			4	6				
	9		3	8		6		
	8		6		2			4
			9		4			5
		5						3
							7	2
	7	4		9				
9	6					5		8

Expert 249

8	1			4				
			8		9	4		1
2					3	6		
						2		
				3				
7								6
		6	7	1				4
1								8
4	3			9	8			7

Expert 250

3	5							
9				6				
1		4	3			9		2
	7					1		6
	6				9			
					5	7		
				1		2	7	5
7	3						1	
		5						3

Expert 251

7				1		4		
				6			7	
1			3			5		
		3		9				1
		9		2				
	1		6	3				5
4					9		2	
	9			5	2	7		

Expert 252

4				6	8			2
	8		9	5				
		6					1	
6			2		9			
9								
		1	7					8
			1		2			
				4		7		
8	2			9	5	4		

Expert 253

4		6		8			7	9
				6				2
					9	1		
9				1				
7			9		2			
	3			4		6		
			1					
8	7	1			4		2	
2			8	3				

Expert 254

			5		9	2		
9				7				1
				3			8	
7		2					6	
5	6					9	3	
	1							7
			6	7		9		
			9	1				3
			2					6

Expert 255

5		8	3		9			2
							3	
			2	4	5			
	1		8			9		
		5						
		3				1	2	
			9	4				3
4	9							
6			2					8

Expert 256

	6			2			3	
				6				
								1
1			6	9				2
		2	3					
	3	5	1	7			6	4
	9			4	2			3
			9		7			
	7	3		4				

Expert 257

			6			9		
			2			4		
		4						
		1				9	2	8
9	6			3	8	1		7
2		7	3					
8	3				4			
4		6	9			8	3	

Expert 258

			6		2			
3						1	9	
		4	7			5	2	3
						4		5
			4	2	5			
	6				9		7	
		1			4	3		
			2					
	9			6				

Expert 259

					7	5	3	
2				4	3	6		
		9						
			2	1	4			
	5							4
	7		6				9	
			3			4		
			2					1
4				5				

Expert 260

6			4	2		8		
				7				6
		2	3				5	4
	5					7		
4	2			6	5			
		7						
3	7			9				
	6				3		4	

Expert 261

			5			9	8	4
				1			7	5
	5	4	3					
				9				3
1								2
	4	8						1
	1		2			3		9
9		3			7		5	

Expert 262

	8		5	4		7		
		2						9
	5	9			3			
1			3		5		8	
				1			7	
	4	5		2	8			
	3		8					
2						1		
						9		

Expert 263

6				3				8
						4		
7		8	4	1		2		
4			3					6
				6		5	1	
			8					
8			7	6				2
		2		8		4		3
			7					

Expert 264

	8	3					6	1
2			7	9				
				1		4		
				7				
			6			1	8	
								3
	2		1					8
6				2				
	9			8		7		

Expert 265

	6			2	5	9		
			9			1	6	
			8					
	3	4	1				5	
			4		8			1
				6	3			
		3	5					6
5						2	4	
6					9			

Expert 266

1		5		6				7
	7				5			
2			3					
7			6	2			1	
6			5		3			8
			7	8			9	
5	2				1		7	
8							3	
	1					8		

Expert 267

	9		3			6		5
			7	5	8			
1						2		
			7				6	
8		1			6			9
			6			1		
		5			9			
6			5	2	1	9		8

Expert 268

	5		7					
	8	2		4			6	7
6	9		1		2			
	6		9			1		
			6	5		4		
7								
				1		9		
			2	8				
		1				6		8

Expert 269

				6				
5		1	8			4		
	3	8						
	6				8			
2			1		5			3
			4			9	5	
8	1		6		2			
4			5					2
					1			

Expert 270

7				9	4	5		2
	5							3
1		2	3	5				6
4		9						
	3		4					
				2		9		
	2	7	9			4		
			6					5
6		4		7				

Expert 271

	8	2			1		7	
6								
				5			4	
			8	7			2	
	5			2	9			
	9		4			8		
4					7	2		5
7		8		9	4			
		6						7

Expert 272

1			6			5	8	
	6					9		
			8		4			2
4					1			6
8		2					7	
6			5		7		2	
	5		1		6			
			2	7				1

Expert 273

			9		4			
8				2				
	7		6	5		2		
7				4		8		
			5		9			
4		5	2	8				
							8	7
3	2					4		9
		4	3			6	2	

Expert 274

	3	8			4		6	
2			7					
6	1		8			2		
							8	
						4		
			9	7				
						9	7	2
	4		1					
9	8				6	1		

Expert 275

5				6				
	9		3		4	1		
4			1		5			
			2		1	5		
			6				8	
9				4				
						4		8
1			4		6	3	5	
2	3			8				

Expert 276

8			4					
	5							
7						2		
4				7	5			
9			6			8	4	
6			9			5		7
5	3	8				9	2	
	9		8					6
	6			3				5

Expert 277

					3		4	
			6					7
6			4	7			1	8
8	9				7			1
		7		5	9	8		
		4	3					5
			7	9				
	3	9						
7						4		

Expert 278

						9		
	2		4					
5	8				2		3	
	5					8		
2							6	
	9	3						2
	4		6	7		9		
	7		8					5
	2			3				8

Expert 279

	3		7				5	8
				1				
			6		5	9		7
8		4						
					8	6		1
		5	1					
	8		3					
	5	3			1	8		
	6	1		5	7	3		

Expert 280

		2	9			8		1
6				2				
		8		9	1			
		3	4			9		
7			3			1	6	
	4		8		6		2	
				3	9			
			1					7

Expert 281

2		4				9		
						3		
	5	1						
	7				3			
5					9			
3	1			5				8
			1	7		8	9	
	4			8				
9					2		1	7

Expert 282

			2					
	7		8	4		9		
4			5	1		2		
								8
1	9		4					
	4		2	8				
		1	6	2		5	7	
			7	9		6		
	5						4	9

Expert 283

				5				
1			8					
		5	7		2	8		
			6	4				
		7					3	
	4			3		1	8	
5	8	1			6		7	
	2		3	8				5
6		3	1					

Expert 284

				4	7			
	7						4	6
								5
8	9		7					
2		4		3				
5	6							3
6			9			3		
		8			4		6	9
3				6			2	

Expert 285

	2		8	6		7		5
				1			2	
			7	2		6	8	
								9
	6	8						7
	1	5						8
	7	6			4			1
	9							
4				9				

Expert 286

		1	2					
8			3	1	9			7
	4							
	3	6	4			8		2
2						7		1
9			8		2		7	
			1		6			9
				4		2	3	

Expert 287

	2			5		1		
6	3							
				3	8			
7								
		4			7			3
1				6		8		
					3			
	8	6		1	2	5		7
4		2		8			6	

Expert 288

						6		3
					8	7	4	
		4					2	5
	5							
		7						6
3	8	9				2		
			2			4	8	7
7				6				2
4	3				9			

Expert 289

	4		8		1			9
				9				
			4		2			
7	1	9					6	4
5								
					9		2	5
	5	1		4				
			1		8	2		
		6	7					

Expert 290

		8	7		1	5		2
		5		6			1	
	1					4		
	6							
								7
5		2				6		
	8	3		7		2	4	
			4	8				1
1			2	3		7	6	

Expert 291

5			7	6	3		4	
7		6		4		1	5	
	6		1	5				
8	9			3		4		
					8			
			5		4			
				9			7	
	3			7		9	1	

Expert 292

				6		1		2
	5	4				9		
				5				
	4		2			8		
	8		7				2	6
2		1				5		4
	4	8			6		1	
			1			8	7	

Expert 293

	6	5						
	9			6		4		
1			3	9		6	8	
4			8					
			2		5			
	5		9	4				
3						8		
	2	4	6			3	1	
		1		2			9	

Expert 294

8					3	2		
	2	9				7	6	
		4						
		6	2	3		8		
		9	8		7			
	7			9	1			2
	1	3						
		7	4	6				
								6

Expert 295

				5		3	6	
8			1					2
1	6	5		7				8
	7	4		6	2			
4	5					1	7	
6				3				
	2		7	1				

Expert 296

		8					6	
	5		1	8				2
4						5		
8	2			4				
							7	
						5		8
	8	6	2					
	7	9	5			6		
		2	6			1		9

Expert 297

								6
	9	7						
		8	4					7
			9					
8		6	7		5		2	
7	5		8		2			
3						8	6	
		4	5		8	2		
		5		7	3	4		

Expert 298

	1					2	7	6
	5				6			1
1				4				8
						4		
	4					9	2	
	7		5				9	
	6	5			1	8		2
4	2			6	9			

Expert 299

9					4			
		7		1				8
5		3						
			6					
				5	9			
4	6	5		8	7	1		
		8			3		9	6
6			7				4	
1						7		

Expert 300

	1							7
			8	1				4
8			3		9			
	9		7				8	6
7			6	2		3		1
3	7				4			
4	2						6	
		6			3	4		

Easy Solution 1

```
1 7 4 | 6 8 2 | 5 3 9
8 6 2 | 3 5 9 | 7 4 1
9 5 3 | 7 4 1 | 8 6 2
------+-------+------
4 3 1 | 5 2 8 | 6 9 7
7 9 8 | 1 6 3 | 2 5 4
6 2 5 | 9 7 4 | 3 1 8
------+-------+------
5 1 6 | 2 9 7 | 4 8 3
3 4 7 | 8 1 6 | 9 2 5
2 8 9 | 4 3 5 | 1 7 6
```

Easy Solution 2

```
4 7 8 | 6 9 2 | 3 1 5
2 1 5 | 4 8 3 | 6 7 9
3 9 6 | 7 5 1 | 2 4 8
------+-------+------
1 8 3 | 9 2 5 | 7 6 4
9 6 4 | 8 1 7 | 5 2 3
5 2 7 | 3 4 6 | 9 8 1
------+-------+------
7 5 9 | 1 6 8 | 4 3 2
8 3 2 | 5 7 4 | 1 9 6
6 4 1 | 2 3 9 | 8 5 7
```

Easy Solution 3

```
4 7 1 | 9 3 2 | 5 8 6
2 5 8 | 6 1 4 | 3 7 9
6 3 9 | 7 8 5 | 2 4 1
------+-------+------
8 9 6 | 3 7 1 | 4 2 5
3 1 2 | 4 5 6 | 8 9 7
7 4 5 | 2 9 8 | 1 6 3
------+-------+------
5 2 3 | 8 6 9 | 7 1 4
9 8 7 | 1 4 3 | 6 5 2
1 6 4 | 5 2 7 | 9 3 8
```

Easy Solution 4

```
2 1 7 | 9 4 3 | 8 5 6
3 5 4 | 8 6 1 | 2 9 7
8 9 6 | 2 5 7 | 1 4 3
------+-------+------
4 2 1 | 7 3 6 | 9 8 5
7 8 5 | 4 2 9 | 3 6 1
6 3 9 | 1 8 5 | 7 2 4
------+-------+------
9 4 3 | 5 7 8 | 6 1 2
5 6 8 | 3 1 2 | 4 7 9
1 7 2 | 6 9 4 | 5 3 8
```

Easy Solution 5

```
1 7 3 | 9 6 8 | 5 2 4
6 2 4 | 7 3 5 | 9 1 8
8 5 9 | 2 4 1 | 6 7 3
------+-------+------
7 6 8 | 3 9 2 | 4 5 1
2 3 1 | 5 7 4 | 8 6 9
4 9 5 | 1 8 6 | 2 3 7
------+-------+------
3 8 2 | 4 5 7 | 1 9 6
5 4 7 | 6 1 9 | 3 8 2
9 1 6 | 8 2 3 | 7 4 5
```

Easy Solution 6

```
6 4 3 | 1 9 7 | 8 5 2
7 5 9 | 2 6 8 | 1 3 4
2 8 1 | 4 3 5 | 7 6 9
------+-------+------
9 1 6 | 7 8 4 | 5 2 3
8 2 7 | 3 5 9 | 4 1 6
5 3 4 | 6 2 1 | 9 8 7
------+-------+------
4 6 8 | 9 1 3 | 2 7 5
3 7 5 | 8 4 2 | 6 9 1
1 9 2 | 5 7 6 | 3 4 8
```

Easy Solution 7

```
6 4 8 | 1 7 5 | 9 2 3
5 9 7 | 3 8 2 | 4 1 6
2 1 3 | 6 4 9 | 8 7 5
------+-------+------
3 6 1 | 7 5 8 | 2 9 4
7 5 9 | 2 6 4 | 1 3 8
8 2 4 | 9 3 1 | 6 5 7
------+-------+------
4 3 2 | 5 1 6 | 7 8 9
1 8 5 | 4 9 7 | 3 6 2
9 7 6 | 8 2 3 | 5 4 1
```

Easy Solution 8

```
1 3 6 | 4 2 7 | 9 8 5
9 8 5 | 3 6 1 | 7 2 4
4 7 2 | 5 9 8 | 1 6 3
------+-------+------
2 6 8 | 1 4 3 | 5 7 9
3 9 4 | 7 5 2 | 8 1 6
5 1 7 | 9 8 6 | 3 4 2
------+-------+------
6 2 3 | 8 7 9 | 4 5 1
7 4 9 | 6 1 5 | 2 3 8
8 5 1 | 2 3 4 | 6 9 7
```

Easy Solution 9

```
5 1 3 | 8 9 4 | 2 6 7
2 9 8 | 7 5 6 | 1 3 4
4 7 6 | 3 1 2 | 9 5 8
------+-------+------
7 4 5 | 6 8 1 | 3 9 2
3 6 2 | 5 7 9 | 4 8 1
9 8 1 | 4 2 3 | 5 7 6
------+-------+------
6 2 7 | 1 3 5 | 8 4 9
1 3 4 | 9 6 8 | 7 2 5
8 5 9 | 2 4 7 | 6 1 3
```

Easy Solution 10

```
5 6 4 | 1 9 7 | 8 3 2
2 7 9 | 5 3 8 | 6 4 1
3 1 8 | 2 4 6 | 9 7 5
------+-------+------
1 4 3 | 6 8 9 | 2 5 7
7 9 2 | 3 1 5 | 4 8 6
6 8 5 | 7 2 4 | 1 9 3
------+-------+------
4 2 1 | 9 5 3 | 7 6 8
8 5 7 | 4 6 1 | 3 2 9
9 3 6 | 8 7 2 | 5 1 4
```

Easy Solution 11

```
7 9 1 | 3 5 4 | 8 6 2
6 8 2 | 1 9 7 | 4 3 5
5 4 3 | 2 8 6 | 9 7 1
------+-------+------
2 3 7 | 6 1 8 | 5 9 4
4 6 8 | 9 2 5 | 7 1 3
9 1 5 | 7 4 3 | 2 8 6
------+-------+------
1 7 9 | 4 3 2 | 6 5 8
3 5 4 | 8 6 9 | 1 2 7
8 2 6 | 5 7 1 | 3 4 9
```

Easy Solution 12

```
6 7 5 | 2 4 9 | 8 3 1
2 4 8 | 1 5 3 | 6 9 7
3 9 1 | 8 6 7 | 2 5 4
------+-------+------
4 3 9 | 5 1 6 | 7 2 8
7 1 2 | 9 8 4 | 5 6 3
8 5 6 | 3 7 2 | 1 4 9
------+-------+------
9 8 7 | 4 2 5 | 3 1 6
5 6 4 | 7 3 1 | 9 8 2
1 2 3 | 6 9 8 | 4 7 5
```

Easy Solution 13

```
7 3 4 | 8 5 1 | 9 6 2
1 9 6 | 3 2 7 | 5 8 4
8 5 2 | 6 4 9 | 1 7 3
------+-------+------
5 4 1 | 9 3 8 | 6 2 7
9 7 3 | 5 6 2 | 8 4 1
6 2 8 | 1 7 4 | 3 5 9
------+-------+------
4 8 5 | 2 1 3 | 7 9 6
2 1 9 | 7 8 6 | 4 3 5
3 6 7 | 4 9 5 | 2 1 8
```

Easy Solution 14

```
4 1 2 | 9 3 5 | 6 7 8
7 5 9 | 4 6 8 | 3 2 1
3 6 8 | 2 1 7 | 4 9 5
------+-------+------
5 2 6 | 1 7 4 | 9 8 3
1 9 4 | 3 8 2 | 5 6 7
8 3 7 | 6 5 9 | 2 1 4
------+-------+------
6 4 3 | 8 2 1 | 7 5 9
2 8 5 | 7 9 3 | 1 4 6
9 7 1 | 5 4 6 | 8 3 2
```

Easy Solution 15

```
2 6 8 | 3 1 5 | 7 9 4
9 1 4 | 8 7 6 | 2 3 5
7 5 3 | 9 4 2 | 8 6 1
------+-------+------
1 4 6 | 5 2 3 | 9 7 8
8 7 2 | 1 9 4 | 3 5 6
3 9 5 | 7 6 8 | 1 4 2
------+-------+------
4 2 7 | 6 8 9 | 5 1 3
5 8 1 | 4 3 7 | 6 2 9
6 3 9 | 2 5 1 | 4 8 7
```

Easy Solution 16

```
2 5 7 | 9 8 1 | 3 4 6
3 1 4 | 6 5 2 | 8 7 9
9 8 6 | 4 3 7 | 1 5 2
------+-------+------
5 2 1 | 7 6 3 | 9 8 4
6 3 8 | 2 9 4 | 7 1 5
4 7 9 | 8 1 5 | 2 6 3
------+-------+------
7 6 3 | 5 2 8 | 4 9 1
8 9 2 | 1 4 6 | 5 3 7
1 4 5 | 3 7 9 | 6 2 8
```

Easy Solution 17

```
5 9 7 | 2 6 4 | 3 8 1
2 6 4 | 1 8 3 | 5 7 9
8 3 1 | 7 9 5 | 2 6 4
------+-------+------
7 1 9 | 5 3 2 | 6 4 8
4 5 8 | 6 1 7 | 9 3 2
3 2 6 | 9 4 8 | 7 1 5
------+-------+------
1 4 5 | 3 2 6 | 8 9 7
6 8 2 | 4 7 9 | 1 5 3
9 7 3 | 8 5 1 | 4 2 6
```

Easy Solution 18

```
1 5 2 | 6 8 3 | 7 9 4
3 7 4 | 5 9 1 | 2 6 8
8 9 6 | 7 4 2 | 1 5 3
------+-------+------
5 8 7 | 9 2 4 | 6 3 1
4 2 9 | 3 1 6 | 8 7 5
6 1 3 | 8 5 7 | 9 4 2
------+-------+------
9 4 8 | 1 7 5 | 3 2 6
7 3 5 | 2 6 8 | 4 1 9
2 6 1 | 4 3 9 | 5 8 7
```

Easy Solution 19

```
2 8 7 | 3 9 1 | 6 5 4
1 9 3 | 6 4 5 | 8 2 7
6 4 5 | 7 8 2 | 1 9 3
------+-------+------
7 1 4 | 5 3 9 | 2 8 6
5 2 8 | 4 1 6 | 3 7 9
9 3 6 | 2 7 8 | 5 4 1
------+-------+------
4 5 1 | 9 2 3 | 7 6 8
8 6 9 | 1 5 7 | 4 3 2
3 7 2 | 8 6 4 | 9 1 5
```

Easy Solution 20

```
4 5 3 | 6 2 8 | 7 9 1
8 1 2 | 4 9 7 | 5 6 3
9 7 6 | 1 3 5 | 4 8 2
------+-------+------
1 4 9 | 5 7 3 | 8 2 6
6 3 8 | 2 1 4 | 9 5 7
5 2 7 | 8 6 9 | 1 3 4
------+-------+------
7 9 1 | 3 8 2 | 6 4 5
2 6 5 | 9 4 1 | 3 7 8
3 8 4 | 7 5 6 | 2 1 9
```

Easy Solution 21

7	9	1	5	8	3	6	2	4
6	3	2	1	9	4	5	8	7
4	5	8	7	6	2	3	9	1
3	2	6	9	7	1	4	5	8
9	1	5	6	4	8	2	7	3
8	7	4	2	3	5	1	6	9
5	4	7	3	2	9	8	1	6
2	6	3	8	1	7	9	4	5
1	8	9	4	5	6	7	3	2

Easy Solution 22

2	1	7	8	4	5	3	9	6
5	3	6	1	9	7	8	2	4
4	8	9	2	3	6	7	1	5
6	2	8	7	5	9	1	4	3
1	9	3	4	6	8	2	5	7
7	4	5	3	1	2	6	8	9
9	7	1	6	8	4	5	3	2
3	6	4	5	2	1	9	7	8
8	5	2	9	7	3	4	6	1

Easy Solution 23

2	3	4	8	6	5	7	1	9
5	1	8	9	2	7	4	3	6
7	6	9	1	4	3	2	8	5
8	4	1	6	7	9	3	5	2
6	9	2	5	3	8	1	7	4
3	5	7	2	1	4	9	6	8
1	8	3	4	5	2	6	9	7
4	7	5	3	9	6	8	2	1
9	2	6	7	8	1	5	4	3

Easy Solution 24

2	4	8	3	7	9	1	5	6
9	5	1	8	2	6	7	4	3
6	7	3	5	4	1	8	9	2
7	1	4	6	9	5	2	3	8
5	8	9	2	3	7	4	6	1
3	6	2	1	8	4	5	7	9
1	9	6	7	5	8	3	2	4
8	3	7	4	6	2	9	1	5
4	2	5	9	1	3	6	8	7

Easy Solution 25

2	9	7	6	3	4	1	5	8
6	5	8	9	1	7	3	4	2
1	3	4	5	8	2	9	7	6
8	4	9	3	7	5	2	6	1
7	2	3	1	6	8	5	9	4
5	6	1	4	2	9	8	3	7
4	8	5	2	9	6	7	1	3
3	7	6	8	5	1	4	2	9
9	1	2	7	4	3	6	8	5

Easy Solution 26

3	1	9	6	2	7	4	5	8
2	6	4	9	8	5	3	1	7
5	7	8	1	4	3	6	9	2
7	8	1	5	3	6	9	2	4
9	4	2	8	7	1	5	6	3
6	3	5	2	9	4	8	7	1
1	5	3	7	6	8	2	4	9
8	9	6	4	1	2	7	3	5
4	2	7	3	5	9	1	8	6

Easy Solution 27

5	8	2	1	9	7	3	6	4
6	9	1	8	3	4	5	2	7
3	4	7	2	5	6	1	9	8
1	6	9	4	2	5	7	8	3
2	7	8	6	1	3	9	4	5
4	5	3	7	8	9	6	1	2
9	3	4	5	6	8	2	7	1
7	1	6	3	4	2	8	5	9
8	2	5	9	7	1	4	3	6

Easy Solution 28

9	4	5	2	7	8	3	1	6
6	1	8	4	5	3	2	7	9
7	2	3	6	1	9	5	4	8
3	9	2	1	4	7	6	8	5
8	7	1	5	3	6	4	9	2
5	6	4	9	8	2	1	3	7
4	8	7	3	6	5	9	2	1
1	5	9	8	2	4	7	6	3
2	3	6	7	9	1	8	5	4

Easy Solution 29

4	7	1	8	3	5	9	2	6
9	5	2	6	1	4	7	3	8
6	8	3	9	2	7	5	1	4
8	9	5	1	7	2	4	6	3
1	6	4	3	5	8	2	9	7
3	2	7	4	6	9	1	8	5
2	4	9	5	8	6	3	7	1
5	1	6	7	9	3	8	4	2
7	3	8	2	4	1	6	5	9

Easy Solution 30

6	1	8	3	9	2	4	7	5
7	5	2	6	4	1	8	9	3
9	3	4	5	8	7	2	1	6
1	7	9	4	2	5	6	3	8
4	2	3	8	6	9	7	5	1
5	8	6	1	7	3	9	4	2
3	6	7	2	1	4	5	8	9
8	9	5	7	3	6	1	2	4
2	4	1	9	5	8	3	6	7

Easy Solution 31

5	1	2	6	4	8	3	9	7
8	7	3	5	9	2	4	6	1
4	9	6	3	7	1	8	5	2
3	4	7	2	8	5	6	1	9
1	5	8	4	6	9	2	7	3
2	6	9	1	3	7	5	4	8
7	8	4	9	2	6	1	3	5
6	2	5	7	1	3	9	8	4
9	3	1	8	5	4	7	2	6

Easy Solution 32

7	4	1	9	8	6	2	5	3
8	9	3	4	5	2	6	1	7
2	5	6	3	7	1	9	4	8
9	6	2	5	4	3	7	8	1
3	1	4	8	2	7	5	6	9
5	7	8	6	1	9	3	2	4
1	2	5	7	9	8	4	3	6
6	8	9	2	3	4	1	7	5
4	3	7	1	6	5	8	9	2

Easy Solution 33

6	2	4	1	7	9	3	5	8
8	9	5	4	3	2	7	6	1
3	1	7	8	6	5	2	4	9
2	8	3	9	4	7	6	1	5
7	4	1	6	5	3	9	8	2
5	6	9	2	1	8	4	7	3
1	5	6	3	9	4	8	2	7
4	3	8	7	2	1	5	9	6
9	7	2	5	8	6	1	3	4

Easy Solution 34

5	1	8	2	6	3	9	4	7
4	2	6	5	7	9	3	1	8
3	7	9	8	1	4	5	2	6
9	6	4	1	5	8	2	7	3
8	5	2	4	3	7	6	9	1
1	3	7	9	2	6	4	8	5
2	9	3	6	8	1	7	5	4
7	4	1	3	9	5	8	6	2
6	8	5	7	4	2	1	3	9

Easy Solution 35

4	2	5	1	3	9	8	6	7
7	1	9	4	6	8	3	2	5
3	8	6	2	7	5	1	9	4
2	5	8	6	4	1	7	3	9
6	4	1	7	9	3	5	8	2
9	7	3	5	8	2	4	1	6
8	3	7	9	5	6	2	4	1
5	9	2	3	1	4	6	7	8
1	6	4	8	2	7	9	5	3

Easy Solution 36

8	4	5	2	6	9	1	7	3
7	3	6	5	8	1	2	9	4
1	2	9	3	4	7	8	6	5
2	9	7	4	1	5	6	3	8
3	1	8	9	2	6	5	4	7
6	5	4	7	3	8	9	2	1
9	8	1	6	7	4	3	5	2
5	7	3	8	9	2	4	1	6
4	6	2	1	5	3	7	8	9

Easy Solution 37

5	2	1	4	6	8	3	9	7
7	8	9	2	3	1	6	4	5
3	4	6	5	7	9	8	1	2
1	9	2	3	8	6	5	7	4
4	7	8	1	9	5	2	6	3
6	3	5	7	4	2	1	8	9
9	1	3	8	5	4	7	2	6
8	5	4	6	2	7	9	3	1
2	6	7	9	1	3	4	5	8

Easy Solution 38

8	2	7	5	4	1	3	6	9
9	6	1	2	3	8	5	4	7
5	4	3	9	6	7	1	2	8
6	5	2	1	7	3	8	9	4
3	7	4	8	9	2	6	5	1
1	8	9	4	5	6	2	7	3
7	3	8	6	2	4	9	1	5
2	1	5	7	8	9	4	3	6
4	9	6	3	1	5	7	8	2

Easy Solution 39

4	7	8	9	6	2	3	5	1
9	3	5	1	4	7	2	8	6
1	2	6	3	5	8	7	9	4
2	6	3	5	7	9	4	1	8
8	4	7	2	1	6	9	3	5
5	1	9	4	8	3	6	7	2
3	5	2	8	9	4	1	6	7
7	8	4	6	3	1	5	2	9
6	9	1	7	2	5	8	4	3

Easy Solution 40

7	5	1	4	3	9	2	6	8
3	4	9	2	8	6	1	5	7
2	6	8	7	1	5	4	3	9
1	7	3	6	9	4	5	8	2
8	9	5	1	2	3	6	7	4
4	2	6	8	5	7	3	9	1
5	3	7	9	4	1	8	2	6
9	8	4	3	6	2	7	1	5
6	1	2	5	7	8	9	4	3

Easy Solution 41

2	8	7	1	5	4	3	6	9
5	9	3	6	7	8	4	2	1
1	6	4	3	9	2	5	8	7
4	3	6	2	1	7	9	5	8
9	2	8	4	6	5	7	1	3
7	5	1	8	3	9	6	4	2
3	4	2	9	8	6	1	7	5
8	1	5	7	4	3	2	9	6
6	7	9	5	2	1	8	3	4

Easy Solution 42

6	2	1	7	8	3	5	9	4
5	9	4	2	1	6	7	3	8
8	3	7	9	4	5	6	2	1
9	1	6	5	3	4	2	8	7
2	7	5	1	9	8	4	6	3
3	4	8	6	7	2	1	5	9
4	6	9	3	2	1	8	7	5
7	8	2	4	5	9	3	1	6
1	5	3	8	6	7	9	4	2

Easy Solution 43

7	1	3	9	4	8	5	2	6
9	8	2	5	3	6	4	1	7
5	6	4	7	1	2	9	8	3
2	4	5	6	9	1	3	7	8
8	3	9	2	7	5	6	4	1
6	7	1	3	8	4	2	9	5
3	5	8	4	2	7	1	6	9
4	9	7	1	6	3	8	5	2
1	2	6	8	5	9	7	3	4

Easy Solution 44

8	9	7	4	3	6	1	5	2
4	6	5	2	7	1	9	3	8
2	1	3	8	5	9	7	4	6
5	3	1	6	9	4	2	8	7
7	8	2	3	1	5	6	9	4
6	4	9	7	8	2	5	1	3
9	2	6	5	4	3	8	7	1
3	5	8	1	6	7	4	2	9
1	7	4	9	2	8	3	6	5

Easy Solution 45

5	8	1	3	2	4	9	6	7
7	3	9	6	8	1	2	4	5
2	6	4	9	7	5	1	8	3
8	7	6	2	1	3	4	5	9
1	5	3	7	4	9	6	2	8
4	9	2	5	6	8	7	3	1
9	4	7	8	5	6	3	1	2
6	2	5	1	3	7	8	9	4
3	1	8	4	9	2	5	7	6

Easy Solution 46

7	9	1	6	5	2	8	4	3
3	4	2	1	8	7	6	5	9
5	8	6	4	9	3	2	7	1
2	6	7	3	4	5	1	9	8
8	5	9	2	7	1	3	6	4
1	3	4	8	6	9	5	2	7
9	1	5	7	2	8	4	3	6
6	7	3	5	1	4	9	8	2
4	2	8	9	3	6	7	1	5

Easy Solution 47

3	4	1	6	7	5	9	8	2
6	5	7	8	9	2	4	3	1
8	9	2	1	3	4	6	5	7
7	1	3	4	2	8	5	6	9
5	8	4	9	1	6	7	2	3
2	6	9	3	5	7	8	1	4
9	7	5	2	8	3	1	4	6
1	3	6	5	4	9	2	7	8
4	2	8	7	6	1	3	9	5

Easy Solution 48

8	5	4	2	3	7	6	1	9
6	2	1	8	4	9	5	3	7
7	9	3	1	5	6	2	8	4
5	3	6	4	1	2	7	9	8
9	4	2	6	7	8	3	5	1
1	8	7	3	9	5	4	6	2
3	1	8	5	2	4	9	7	6
4	6	9	7	8	3	1	2	5
2	7	5	9	6	1	8	4	3

Easy Solution 49

5	7	9	8	2	1	4	3	6
4	3	6	7	9	5	8	2	1
1	2	8	4	3	6	7	9	5
8	4	5	2	6	7	9	1	3
7	9	2	5	1	3	6	8	4
6	1	3	9	8	4	2	5	7
9	6	7	1	5	8	3	4	2
3	8	1	6	4	2	5	7	9
2	5	4	3	7	9	1	6	8

Easy Solution 50

7	6	5	9	4	3	8	2	1
9	2	8	7	1	5	3	6	4
1	4	3	6	2	8	5	7	9
8	9	7	1	5	4	2	3	6
6	5	2	8	3	9	4	1	7
4	3	1	2	7	6	9	8	5
3	7	6	5	9	2	1	4	8
2	8	9	4	6	1	7	5	3
5	1	4	3	8	7	6	9	2

Easy Solution 51

3	1	4	6	2	5	7	9	8
9	5	8	7	1	3	4	6	2
7	2	6	9	8	4	3	5	1
2	4	9	3	5	7	1	8	6
5	7	3	8	6	1	2	4	9
8	6	1	2	4	9	5	3	7
4	3	2	1	9	6	8	7	5
1	9	7	5	3	8	6	2	4
6	8	5	4	7	2	9	1	3

Easy Solution 52

8	9	5	2	1	3	4	6	7
6	4	1	7	9	5	3	2	8
2	7	3	6	8	4	1	5	9
4	8	6	5	2	9	7	3	1
7	1	9	4	3	6	2	8	5
5	3	2	8	7	1	9	4	6
3	6	7	9	4	8	5	1	2
9	5	4	1	6	2	8	7	3
1	2	8	3	5	7	6	9	4

Easy Solution 53

1	6	9	5	8	4	2	3	7
3	8	5	2	1	7	6	9	4
7	2	4	6	3	9	1	8	5
5	9	7	8	6	2	3	4	1
8	1	6	4	5	3	9	7	2
2	4	3	9	7	1	8	5	6
9	7	8	1	2	5	4	6	3
4	3	2	7	9	6	5	1	8
6	5	1	3	4	8	7	2	9

Easy Solution 54

5	8	7	3	9	2	6	1	4
3	1	4	5	7	6	2	8	9
9	6	2	4	1	8	3	5	7
7	2	6	8	4	5	1	9	3
8	9	5	1	3	7	4	2	6
1	4	3	6	2	9	5	7	8
6	3	9	2	8	1	7	4	5
4	7	1	9	5	3	8	6	2
2	5	8	7	6	4	9	3	1

Easy Solution 55

5	3	4	2	1	8	9	6	7
8	6	9	4	7	3	2	5	1
2	1	7	9	6	5	3	8	4
6	2	3	1	8	9	7	4	5
1	9	5	7	4	6	8	2	3
4	7	8	3	5	2	6	1	9
9	8	6	5	3	1	4	7	2
3	4	1	6	2	7	5	9	8
7	5	2	8	9	4	1	3	6

Easy Solution 56

3	7	8	5	1	6	2	4	9
6	5	1	4	9	2	8	3	7
4	9	2	7	8	3	6	1	5
2	4	9	8	7	5	3	6	1
5	8	6	9	3	1	7	2	4
1	3	7	2	6	4	9	5	8
9	1	5	3	2	7	4	8	6
8	6	3	1	4	9	5	7	2
7	2	4	6	5	8	1	9	3

Easy Solution 57

4	6	9	5	1	2	7	3	8
3	5	8	9	4	7	6	1	2
1	7	2	8	6	3	5	4	9
8	9	7	3	2	6	4	5	1
2	3	6	4	5	1	8	9	7
5	4	1	7	8	9	3	2	6
6	2	3	1	7	4	9	8	5
7	8	4	2	9	5	1	6	3
9	1	5	6	3	8	2	7	4

Easy Solution 58

1	9	4	2	7	8	5	3	6
7	2	5	3	6	9	1	4	8
8	6	3	1	4	5	2	7	9
3	7	9	8	2	4	6	1	5
5	1	2	6	3	7	9	8	4
4	8	6	5	9	1	3	2	7
6	4	7	9	1	3	8	5	2
2	5	1	4	8	6	7	9	3
9	3	8	7	5	2	4	6	1

Easy Solution 59

4	8	5	9	3	6	2	7	1
3	2	1	5	4	7	9	8	6
7	9	6	1	2	8	5	3	4
2	3	8	6	9	4	7	1	5
6	1	4	3	7	5	8	9	2
9	5	7	8	1	2	4	6	3
1	6	2	7	5	9	3	4	8
5	7	3	4	8	1	6	2	9
8	4	9	2	6	3	1	5	7

Easy Solution 60

2	5	7	6	4	9	8	1	3
9	1	6	8	3	5	4	7	2
4	3	8	1	2	7	5	6	9
7	8	5	2	1	6	3	9	4
6	2	1	3	9	4	7	8	5
3	4	9	7	5	8	1	2	6
8	7	4	9	6	3	2	5	1
5	9	2	4	7	1	6	3	8
1	6	3	5	8	2	9	4	7

Easy Solution 61

9	5	8	7	4	3	6	1	2
3	6	2	9	1	8	7	4	5
7	4	1	6	5	2	3	9	8
8	9	4	3	7	5	2	6	1
2	1	3	4	8	6	5	7	9
6	7	5	2	9	1	8	3	4
4	8	6	5	3	9	1	2	7
5	2	7	1	6	4	9	8	3
1	3	9	8	2	7	4	5	6

Easy Solution 62

5	2	1	8	7	4	6	3	9
9	3	4	2	5	6	7	1	8
6	7	8	1	9	3	5	2	4
8	4	6	3	2	7	1	9	5
1	9	7	4	6	5	2	8	3
2	5	3	9	1	8	4	6	7
4	6	9	7	8	1	3	5	2
7	1	2	5	3	9	8	4	6
3	8	5	6	4	2	9	7	1

Easy Solution 63

3	7	2	5	6	8	1	4	9
4	6	1	7	3	9	5	8	2
8	5	9	2	1	4	6	7	3
1	2	3	8	5	6	4	9	7
6	9	7	3	4	2	8	1	5
5	4	8	9	7	1	3	2	6
7	1	4	6	2	3	9	5	8
2	8	6	1	9	5	7	3	4
9	3	5	4	8	7	2	6	1

Easy Solution 64

1	5	9	3	7	8	2	6	4
2	6	7	9	4	5	1	8	3
4	8	3	2	6	1	7	5	9
3	1	5	4	9	6	8	2	7
7	4	6	8	2	3	5	9	1
9	2	8	1	5	7	4	3	6
6	3	2	5	1	4	9	7	8
8	9	4	7	3	2	6	1	5
5	7	1	6	8	9	3	4	2

Easy Solution 65

4	8	3	1	9	5	7	6	2
5	2	7	6	4	8	1	9	3
1	9	6	7	3	2	5	4	8
2	5	9	8	6	1	3	7	4
6	4	1	3	7	9	8	2	5
7	3	8	2	5	4	9	1	6
3	1	2	4	8	7	6	5	9
8	7	5	9	2	6	4	3	1
9	6	4	5	1	3	2	8	7

Easy Solution 66

5	3	7	2	6	8	1	9	4
6	9	8	4	1	3	5	2	7
1	4	2	9	5	7	6	8	3
9	1	4	3	2	5	7	6	8
8	6	3	1	7	4	9	5	2
7	2	5	8	9	6	4	3	1
4	8	1	5	3	9	2	7	6
2	5	6	7	8	1	3	4	9
3	7	9	6	4	2	8	1	5

Easy Solution 67

2	4	7	6	1	3	8	5	9
5	3	9	2	7	8	6	1	4
1	6	8	9	4	5	2	7	3
6	7	5	8	9	4	3	2	1
8	9	2	3	6	1	5	4	7
3	1	4	7	5	2	9	6	8
9	2	1	4	8	6	7	3	5
7	5	6	1	3	9	4	8	2
4	8	3	5	2	7	1	9	6

Easy Solution 68

5	3	4	7	6	8	2	1	9
1	6	8	2	9	3	5	7	4
2	9	7	1	4	5	8	3	6
4	7	1	3	5	6	9	2	8
3	5	2	9	8	1	6	4	7
6	8	9	4	2	7	1	5	3
8	4	5	6	7	2	3	9	1
7	2	3	8	1	9	4	6	5
9	1	6	5	3	4	7	8	2

Easy Solution 69

9	7	6	4	5	3	8	1	2
1	8	4	9	2	6	7	3	5
2	5	3	8	7	1	9	4	6
5	2	8	3	6	7	4	9	1
6	4	1	2	8	9	3	5	7
3	9	7	1	4	5	2	6	8
7	3	9	5	1	8	6	2	4
8	1	2	6	3	4	5	7	9
4	6	5	7	9	2	1	8	3

Easy Solution 70

6	1	2	5	3	9	4	7	8
5	4	8	2	7	1	6	9	3
7	9	3	8	6	4	5	1	2
4	5	9	1	2	7	8	3	6
2	7	6	3	9	8	1	4	5
8	3	1	4	5	6	7	2	9
3	2	7	6	1	5	9	8	4
9	8	5	7	4	2	3	6	1
1	6	4	9	8	3	2	5	7

Easy Solution 71

3	9	6	1	2	8	5	7	4
1	5	8	7	4	3	2	6	9
4	7	2	9	6	5	8	3	1
6	4	9	8	7	2	1	5	3
8	1	7	3	5	9	6	4	2
5	2	3	4	1	6	9	8	7
7	6	5	2	9	4	3	1	8
9	8	1	5	3	7	4	2	6
2	3	4	6	8	1	7	9	5

Easy Solution 72

8	7	4	5	3	6	1	2	9
3	2	9	7	8	1	5	6	4
6	1	5	2	4	9	3	8	7
5	8	2	4	6	7	9	1	3
9	6	1	3	2	5	4	7	8
4	3	7	9	1	8	6	5	2
1	9	6	8	7	4	2	3	5
2	5	8	1	9	3	7	4	6
7	4	3	6	5	2	8	9	1

Easy Solution 73

8	9	4	3	2	7	6	1	5
5	7	2	9	1	6	4	3	8
3	6	1	5	8	4	2	7	9
6	5	9	1	4	3	8	2	7
2	1	8	7	6	9	5	4	3
7	4	3	8	5	2	9	6	1
9	3	6	2	7	8	1	5	4
1	2	7	4	9	5	3	8	6
4	8	5	6	3	1	7	9	2

Easy Solution 74

4	6	9	1	8	3	7	2	5
3	2	5	6	9	7	1	4	8
7	1	8	4	2	5	6	9	3
2	8	3	7	4	9	5	6	1
1	5	4	8	6	2	9	3	7
6	9	7	5	3	1	4	8	2
9	7	6	3	5	8	2	1	4
8	4	1	2	7	6	3	5	9
5	3	2	9	1	4	8	7	6

Easy Solution 75

5	8	2	7	3	4	6	9	1
1	9	7	5	6	8	4	2	3
6	3	4	9	1	2	8	7	5
8	4	5	1	2	6	9	3	7
9	7	3	4	8	5	1	6	2
2	1	6	3	9	7	5	4	8
4	5	8	6	7	3	2	1	9
3	2	1	8	4	9	7	5	6
7	6	9	2	5	1	3	8	4

Easy Solution 76

7	6	5	4	8	9	2	1	3
1	2	8	3	5	7	9	6	4
3	4	9	1	6	2	7	8	5
2	8	1	5	9	4	6	3	7
4	7	6	2	3	1	8	5	9
9	5	3	8	7	6	1	4	2
8	9	7	6	4	3	5	2	1
5	1	4	9	2	8	3	7	6
6	3	2	7	1	5	4	9	8

Easy Solution 77

1	7	9	3	5	8	2	6	4
2	6	8	7	4	9	3	5	1
3	4	5	2	6	1	9	7	8
5	9	7	1	3	2	4	8	6
6	3	1	4	8	5	7	2	9
8	2	4	6	9	7	1	3	5
9	1	3	5	2	6	8	4	7
7	5	2	8	1	4	6	9	3
4	8	6	9	7	3	5	1	2

Easy Solution 78

8	4	2	9	7	1	6	3	5
6	1	7	5	3	8	4	9	2
5	9	3	6	2	4	7	8	1
4	2	5	3	9	7	1	6	8
7	8	6	1	4	2	9	5	3
9	3	1	8	6	5	2	7	4
2	5	8	7	1	9	3	4	6
3	7	4	2	5	6	8	1	9
1	6	9	4	8	3	5	2	7

Easy Solution 79

8	6	5	4	1	7	9	3	2
3	1	7	9	8	2	4	6	5
4	2	9	3	5	6	7	8	1
1	7	3	5	2	9	8	4	6
6	9	4	8	3	1	5	2	7
2	5	8	7	6	4	3	1	9
9	4	1	6	7	8	2	5	3
5	8	6	2	9	3	1	7	4
7	3	2	1	4	5	6	9	8

Easy Solution 80

8	3	5	2	6	1	7	9	4
1	9	4	7	5	8	3	2	6
2	7	6	4	3	9	1	5	8
6	8	1	9	4	3	2	7	5
3	5	2	8	1	7	4	6	9
9	4	7	5	2	6	8	3	1
4	2	8	3	9	5	6	1	7
5	1	3	6	7	4	9	8	2
7	6	9	1	8	2	5	4	3

Easy Solution 81

7	1	2	4	9	8	6	5	3
9	8	3	2	6	5	1	4	7
6	4	5	1	7	3	9	8	2
8	9	1	5	2	7	4	3	6
5	2	4	9	3	6	8	7	1
3	7	6	8	4	1	2	9	5
1	6	9	3	5	4	7	2	8
4	3	8	7	1	2	5	6	9
2	5	7	6	8	9	3	1	4

Easy Solution 82

2	5	9	6	3	1	4	7	8
6	1	7	8	5	4	3	9	2
3	8	4	2	7	9	6	5	1
9	7	8	4	2	5	1	6	3
1	2	3	9	6	7	8	4	5
5	4	6	1	8	3	9	2	7
7	6	1	3	4	2	5	8	9
4	3	5	7	9	8	2	1	6
8	9	2	5	1	6	7	3	4

Easy Solution 83

9	7	1	6	8	3	4	2	5
6	2	5	4	7	9	1	3	8
8	3	4	2	5	1	6	9	7
5	6	2	9	4	7	8	1	3
1	9	7	8	3	5	2	6	4
3	4	8	1	6	2	7	5	9
4	1	9	3	2	8	5	7	6
2	5	6	7	9	4	3	8	1
7	8	3	5	1	6	9	4	2

Easy Solution 84

9	2	4	1	7	5	6	3	8
7	3	8	4	2	6	1	5	9
1	5	6	8	3	9	7	4	2
5	7	2	6	4	3	9	8	1
3	8	1	9	5	7	4	2	6
6	4	9	2	8	1	3	7	5
4	1	5	7	9	2	8	6	3
8	6	3	5	1	4	2	9	7
2	9	7	3	6	8	5	1	4

Easy Solution 85

2	8	9	1	3	4	6	5	7
3	4	6	8	5	7	1	9	2
7	5	1	9	2	6	8	3	4
8	2	5	4	7	1	9	6	3
6	9	4	3	8	5	7	2	1
1	7	3	6	9	2	5	4	8
5	3	2	7	6	8	4	1	9
9	1	7	5	4	3	2	8	6
4	6	8	2	1	9	3	7	5

Easy Solution 86

8	1	6	7	9	3	5	4	2
5	3	7	4	2	6	8	1	9
2	9	4	1	5	8	6	3	7
1	7	3	5	6	2	4	9	8
6	2	5	8	4	9	3	7	1
9	4	8	3	7	1	2	5	6
7	8	9	6	3	5	1	2	4
3	6	2	9	1	4	7	8	5
4	5	1	2	8	7	9	6	3

Easy Solution 87

5	3	8	4	6	2	9	1	7
1	2	6	9	8	7	3	5	4
9	4	7	5	1	3	6	2	8
2	7	5	3	9	4	1	8	6
3	8	9	6	7	1	2	4	5
4	6	1	2	5	8	7	3	9
6	1	4	8	3	9	5	7	2
8	9	3	7	2	5	4	6	1
7	5	2	1	4	6	8	9	3

Easy Solution 88

7	8	6	9	1	3	5	4	2
3	5	9	6	2	4	8	7	1
2	4	1	7	5	8	9	6	3
4	2	5	8	6	7	1	3	9
9	6	3	1	4	2	7	5	8
1	7	8	3	9	5	4	2	6
8	1	2	5	7	6	3	9	4
6	3	7	4	8	9	2	1	5
5	9	4	2	3	1	6	8	7

Easy Solution 89

2	9	7	5	8	1	3	6	4
8	4	3	2	9	6	7	1	5
5	1	6	3	7	4	8	9	2
6	5	9	4	3	8	2	7	1
4	8	1	9	2	7	6	5	3
7	3	2	6	1	5	4	8	9
9	6	5	8	4	2	1	3	7
1	2	8	7	5	3	9	4	6
3	7	4	1	6	9	5	2	8

Easy Solution 90

9	4	1	5	3	8	2	6	7
5	2	8	7	4	6	1	3	9
6	3	7	9	2	1	4	8	5
2	8	3	1	5	9	7	4	6
1	5	9	4	6	7	8	2	3
4	7	6	3	8	2	5	9	1
8	1	2	6	9	5	3	7	4
7	6	4	8	1	3	9	5	2
3	9	5	2	7	4	6	1	8

Easy Solution 91

5	2	6	4	9	3	8	1	7
8	1	3	2	7	5	6	4	9
7	9	4	1	6	8	3	5	2
6	5	8	9	2	7	4	3	1
3	7	9	5	1	4	2	6	8
1	4	2	3	8	6	7	9	5
2	8	5	6	4	1	9	7	3
9	6	1	7	3	2	5	8	4
4	3	7	8	5	9	1	2	6

Easy Solution 92

6	5	7	8	3	2	4	1	9
3	4	1	7	9	5	2	8	6
2	9	8	6	1	4	3	7	5
8	6	9	4	2	3	1	5	7
7	2	3	5	8	1	6	9	4
5	1	4	9	7	6	8	2	3
1	8	6	3	5	7	9	4	2
4	7	2	1	6	9	5	3	8
9	3	5	2	4	8	7	6	1

Easy Solution 93

4	2	9	8	6	5	7	3	1
8	7	6	9	1	3	2	5	4
5	3	1	2	4	7	8	9	6
6	1	2	4	3	9	5	7	8
3	5	8	6	7	2	1	4	9
9	4	7	1	5	8	6	2	3
7	8	4	3	2	6	9	1	5
1	6	5	7	9	4	3	8	2
2	9	3	5	8	1	4	6	7

Easy Solution 94

4	1	5	3	6	7	9	8	2
9	3	2	4	1	8	7	5	6
8	7	6	2	5	9	3	4	1
2	9	8	5	3	6	1	7	4
6	4	3	7	9	1	8	2	5
7	5	1	8	2	4	6	3	9
5	2	9	1	8	3	4	6	7
1	8	7	6	4	2	5	9	3
3	6	4	9	7	5	2	1	8

Easy Solution 95

9	6	7	4	5	2	1	3	8
8	2	1	3	9	6	7	5	4
3	5	4	8	7	1	9	6	2
2	4	3	5	8	7	6	9	1
7	9	6	1	2	4	3	8	5
5	1	8	9	6	3	2	4	7
1	3	9	7	4	8	5	2	6
4	7	2	6	3	5	8	1	9
6	8	5	2	1	9	4	7	3

Easy Solution 96

7	2	3	1	4	6	9	8	5
4	9	5	8	3	2	1	6	7
8	6	1	9	7	5	4	2	3
3	8	4	7	2	9	5	1	6
5	7	6	3	1	4	8	9	2
9	1	2	6	5	8	7	3	4
6	3	9	4	8	7	2	5	1
1	5	7	2	9	3	6	4	8
2	4	8	5	6	1	3	7	9

Easy Solution 97

2	7	5	3	4	9	8	6	1
9	3	1	7	6	8	4	2	5
8	6	4	5	2	1	7	3	9
7	5	9	6	8	2	1	4	3
3	4	6	1	9	5	2	7	8
1	2	8	4	3	7	9	5	6
6	8	3	2	1	4	5	9	7
4	1	7	9	5	6	3	8	2
5	9	2	8	7	3	6	1	4

Easy Solution 98

8	1	3	7	5	9	6	2	4
9	2	4	8	6	1	7	3	5
7	5	6	3	4	2	8	1	9
4	8	9	1	3	5	2	7	6
2	6	5	9	8	7	1	4	3
3	7	1	6	2	4	9	5	8
5	9	2	4	7	8	3	6	1
6	4	8	2	1	3	5	9	7
1	3	7	5	9	6	4	8	2

Easy Solution 99

8	1	2	7	6	3	9	5	4
6	9	3	8	4	5	2	1	7
4	5	7	1	9	2	6	8	3
3	2	9	6	1	8	4	7	5
5	8	1	2	7	4	3	9	6
7	4	6	5	3	9	8	2	1
1	7	8	3	2	6	5	4	9
2	6	4	9	5	1	7	3	8
9	3	5	4	8	7	1	6	2

Easy Solution 100

6	5	1	9	2	7	4	3	8
2	3	7	6	4	8	1	5	9
9	8	4	3	5	1	7	2	6
8	6	9	7	3	4	2	1	5
7	4	5	8	1	2	9	6	3
1	2	3	5	9	6	8	7	4
3	9	2	4	7	5	6	8	1
5	1	6	2	8	9	3	4	7
4	7	8	1	6	3	5	9	2

Easy Solution 101

2	5	6	9	1	3	8	4	7
1	8	3	6	7	4	5	2	9
9	4	7	2	5	8	1	3	6
3	6	9	8	2	5	4	7	1
8	7	5	3	4	1	6	9	2
4	1	2	7	6	9	3	8	5
6	3	4	1	9	2	7	5	8
7	2	8	5	3	6	9	1	4
5	9	1	4	8	7	2	6	3

Easy Solution 102

5	1	9	8	7	6	3	2	4
8	4	7	2	3	9	6	1	5
2	3	6	1	5	4	8	7	9
9	2	5	7	6	1	4	8	3
3	7	4	5	9	8	2	6	1
6	8	1	3	4	2	9	5	7
7	5	2	4	8	3	1	9	6
4	9	8	6	1	7	5	3	2
1	6	3	9	2	5	7	4	8

Easy Solution 103

9	2	7	6	1	5	3	8	4
1	4	5	3	7	8	6	9	2
6	8	3	2	9	4	1	5	7
2	1	4	9	8	7	5	6	3
3	7	8	5	2	6	9	4	1
5	6	9	4	3	1	7	2	8
8	9	2	1	5	3	4	7	6
4	5	1	7	6	2	8	3	9
7	3	6	8	4	9	2	1	5

Easy Solution 104

9	3	7	5	2	6	1	8	4
1	5	2	4	3	8	9	7	6
6	4	8	7	9	1	5	3	2
8	2	1	6	4	7	3	9	5
4	7	3	8	5	9	2	6	1
5	6	9	2	1	3	8	4	7
7	1	4	3	8	5	6	2	9
3	9	6	1	7	2	4	5	8
2	8	5	9	6	4	7	1	3

Easy Solution 105

6	1	4	7	5	3	8	2	9
7	3	2	9	8	1	5	6	4
5	9	8	2	6	4	3	1	7
4	5	3	6	7	9	2	8	1
2	6	1	8	4	5	7	9	3
9	8	7	3	1	2	6	4	5
8	2	9	4	3	7	1	5	6
1	7	6	5	9	8	4	3	2
3	4	5	1	2	6	9	7	8

Easy Solution 106

7	8	6	1	3	4	2	9	5
9	1	3	5	7	2	4	8	6
4	2	5	6	9	8	7	1	3
3	4	8	7	5	9	1	6	2
1	5	7	2	6	3	9	4	8
2	6	9	4	8	1	5	3	7
6	3	2	9	4	7	8	5	1
5	9	1	8	2	6	3	7	4
8	7	4	3	1	5	6	2	9

Easy Solution 107

6	5	1	7	2	8	3	9	4
8	2	7	3	9	4	5	6	1
9	4	3	1	5	6	2	8	7
7	9	5	4	6	1	8	2	3
3	8	4	5	7	2	6	1	9
1	6	2	9	8	3	4	7	5
2	1	9	6	4	5	7	3	8
4	3	6	8	1	7	9	5	2
5	7	8	2	3	9	1	4	6

Easy Solution 108

1	6	2	9	8	3	7	4	5
4	3	5	2	7	1	8	6	9
7	8	9	5	6	4	1	2	3
2	5	1	4	3	8	6	9	7
3	7	4	1	9	6	2	5	8
8	9	6	7	2	5	4	3	1
5	2	3	6	1	7	9	8	4
6	4	7	8	5	9	3	1	2
9	1	8	3	4	2	5	7	6

Easy Solution 109

3	7	2	4	5	1	8	9	6
1	6	9	8	2	7	3	5	4
5	4	8	6	3	9	1	2	7
9	5	1	2	7	3	4	6	8
8	2	6	5	1	4	7	3	9
7	3	4	9	8	6	2	1	5
4	8	3	1	9	5	6	7	2
6	1	5	7	4	2	9	8	3
2	9	7	3	6	8	5	4	1

Easy Solution 110

7	8	1	4	5	2	9	3	6
9	2	4	1	3	6	7	8	5
6	5	3	9	8	7	4	2	1
1	4	6	7	2	9	8	5	3
2	9	8	3	1	5	6	7	4
5	3	7	8	6	4	2	1	9
4	1	9	2	7	3	5	6	8
3	7	5	6	4	8	1	9	2
8	6	2	5	9	1	3	4	7

Easy Solution 111

5	3	7	9	4	1	2	8	6
9	6	8	3	7	2	1	5	4
2	4	1	6	5	8	3	9	7
7	2	3	8	9	4	5	6	1
8	9	4	5	1	6	7	2	3
6	1	5	7	2	3	8	4	9
4	5	9	2	3	7	6	1	8
1	7	6	4	8	5	9	3	2
3	8	2	1	6	9	4	7	5

Easy Solution 112

5	9	8	2	6	1	4	7	3
3	2	4	7	5	9	8	6	1
7	6	1	8	3	4	5	2	9
4	7	2	3	9	5	1	8	6
9	8	5	1	7	6	2	3	4
1	3	6	4	8	2	7	9	5
2	5	7	6	1	3	9	4	8
8	1	3	9	4	7	6	5	2
6	4	9	5	2	8	3	1	7

Easy Solution 113

6	1	9	8	7	3	2	5	4
8	7	3	2	4	5	6	9	1
4	5	2	6	9	1	7	8	3
9	4	1	3	6	8	5	2	7
2	8	6	4	5	7	1	3	9
7	3	5	9	1	2	8	4	6
3	6	7	5	8	9	4	1	2
5	2	4	1	3	6	9	7	8
1	9	8	7	2	4	3	6	5

Easy Solution 114

1	6	7	9	5	3	8	4	2
3	5	4	6	8	2	1	9	7
9	8	2	1	4	7	6	5	3
8	7	9	4	3	6	2	1	5
5	4	6	2	7	1	3	8	9
2	3	1	8	9	5	7	6	4
7	9	3	5	1	8	4	2	6
4	2	8	3	6	9	5	7	1
6	1	5	7	2	4	9	3	8

Easy Solution 115

3	1	7	8	5	9	6	4	2
6	2	8	1	7	4	3	5	9
4	5	9	6	3	2	7	1	8
8	9	2	3	4	5	1	7	6
1	3	4	9	6	7	2	8	5
7	6	5	2	1	8	9	3	4
2	7	1	4	8	6	5	9	3
5	8	6	7	9	3	4	2	1
9	4	3	5	2	1	8	6	7

Easy Solution 116

6	1	4	9	2	5	8	3	7
2	9	5	3	7	8	6	1	4
8	7	3	6	4	1	9	2	5
4	8	7	1	9	6	3	5	2
3	5	6	4	8	2	1	7	9
9	2	1	7	5	3	4	8	6
1	6	9	5	3	7	2	4	8
5	4	2	8	1	9	7	6	3
7	3	8	2	6	4	5	9	1

Easy Solution 117

1	5	9	4	7	3	8	2	6
8	6	7	1	2	5	9	3	4
4	2	3	8	9	6	1	5	7
9	1	6	7	3	2	5	4	8
2	3	4	5	6	8	7	9	1
5	7	8	9	4	1	3	6	2
3	4	5	6	1	7	2	8	9
7	9	2	3	8	4	6	1	5
6	8	1	2	5	9	4	7	3

Easy Solution 118

3	4	2	8	5	6	9	1	7
7	9	6	4	1	2	8	5	3
5	8	1	3	7	9	6	2	4
8	1	3	9	6	7	5	4	2
2	5	9	1	8	4	7	3	6
6	7	4	5	2	3	1	8	9
9	3	5	6	4	1	2	7	8
1	6	7	2	3	8	4	9	5
4	2	8	7	9	5	3	6	1

Easy Solution 119

7	2	8	4	5	1	9	6	3
9	1	3	6	2	7	4	8	5
6	5	4	3	8	9	7	2	1
8	3	5	9	1	2	6	4	7
4	7	1	5	6	8	3	9	2
2	9	6	7	3	4	1	5	8
5	4	9	8	7	3	2	1	6
3	6	2	1	9	5	8	7	4
1	8	7	2	4	6	5	3	9

Easy Solution 120

8	3	5	6	1	9	4	2	7
7	4	1	8	3	2	6	9	5
9	2	6	4	7	5	3	1	8
4	7	2	3	9	8	5	6	1
6	1	3	5	2	4	8	7	9
5	8	9	1	6	7	2	4	3
1	5	7	2	4	3	9	8	6
2	9	8	7	5	6	1	3	4
3	6	4	9	8	1	7	5	2

Easy Solution 121

6	4	3	2	9	1	5	8	7
2	8	5	3	7	4	1	9	6
9	7	1	6	5	8	2	3	4
5	2	9	1	4	7	3	6	8
3	1	7	8	6	2	4	5	9
4	6	8	9	3	5	7	1	2
1	9	4	7	8	3	6	2	5
8	5	2	4	1	6	9	7	3
7	3	6	5	2	9	8	4	1

Easy Solution 122

6	3	5	9	4	7	8	1	2
8	4	9	6	1	2	7	5	3
2	1	7	5	3	8	4	6	9
1	6	4	7	8	9	3	2	5
3	5	8	4	2	1	9	7	6
9	7	2	3	5	6	1	8	4
5	8	1	2	9	4	6	3	7
4	2	6	8	7	3	5	9	1
7	9	3	1	6	5	2	4	8

Easy Solution 123

9	2	5	1	6	7	8	3	4
8	4	7	3	2	9	6	1	5
1	3	6	4	5	8	9	2	7
2	9	4	5	1	3	7	8	6
3	5	1	7	8	6	4	9	2
6	7	8	9	4	2	1	5	3
4	1	3	6	9	5	2	7	8
5	8	9	2	7	4	3	6	1
7	6	2	8	3	1	5	4	9

Easy Solution 124

9	3	4	7	5	1	6	2	8
2	8	1	3	6	9	7	5	4
5	7	6	4	2	8	9	3	1
7	4	9	1	3	6	2	8	5
8	5	3	9	4	2	1	7	6
1	6	2	5	8	7	4	9	3
6	9	7	8	1	5	3	4	2
4	1	8	2	7	3	5	6	9
3	2	5	6	9	4	8	1	7

Easy Solution 125

8	9	6	7	2	4	1	5	3
3	5	2	8	1	9	6	4	7
7	4	1	5	3	6	8	2	9
1	8	5	6	4	7	9	3	2
2	7	4	3	9	8	5	1	6
6	3	9	1	5	2	4	7	8
4	6	7	2	8	5	3	9	1
5	1	8	9	7	3	2	6	4
9	2	3	4	6	1	7	8	5

Easy Solution 126

2	3	4	1	5	6	7	8	9
7	5	1	9	3	8	6	2	4
8	6	9	7	4	2	5	3	1
5	8	2	3	1	9	4	7	6
1	4	3	6	7	5	8	9	2
6	9	7	2	8	4	1	5	3
4	2	5	8	9	1	3	6	7
9	7	8	4	6	3	2	1	5
3	1	6	5	2	7	9	4	8

Easy Solution 127

4	1	7	2	6	5	9	8	3
6	9	2	8	7	3	1	4	5
5	3	8	4	1	9	7	6	2
8	6	3	5	4	1	2	9	7
7	5	4	9	8	2	3	1	6
1	2	9	6	3	7	4	5	8
3	7	5	1	9	6	8	2	4
2	4	1	3	5	8	6	7	9
9	8	6	7	2	4	5	3	1

Easy Solution 128

8	9	4	1	7	5	2	3	6
7	3	6	8	4	2	1	5	9
5	2	1	3	6	9	4	8	7
9	7	2	5	3	6	8	4	1
3	6	8	4	9	1	5	7	2
1	4	5	7	2	8	9	6	3
2	5	9	6	8	3	7	1	4
4	1	3	2	5	7	6	9	8
6	8	7	9	1	4	3	2	5

Easy Solution 129

8	7	2	6	4	9	1	5	3
5	4	1	3	8	7	6	9	2
9	6	3	1	2	5	8	7	4
2	9	6	8	5	1	4	3	7
7	8	5	4	9	3	2	1	6
3	1	4	2	7	6	5	8	9
4	5	9	7	6	8	3	2	1
6	3	8	9	1	2	7	4	5
1	2	7	5	3	4	9	6	8

Easy Solution 130

5	2	6	9	3	8	4	1	7
3	4	8	7	1	6	9	2	5
7	9	1	2	4	5	8	3	6
4	1	7	3	5	9	6	8	2
9	6	3	8	2	1	7	5	4
8	5	2	6	7	4	1	9	3
6	8	4	5	9	3	2	7	1
2	3	9	1	6	7	5	4	8
1	7	5	4	8	2	3	6	9

Easy Solution 131

7	6	4	2	1	5	8	9	3
5	8	1	9	3	7	6	4	2
3	2	9	4	8	6	5	7	1
1	5	3	7	6	4	2	8	9
9	7	8	3	2	1	4	6	5
6	4	2	8	5	9	3	1	7
4	3	6	1	7	2	9	5	8
8	1	5	6	9	3	7	2	4
2	9	7	5	4	8	1	3	6

Easy Solution 132

9	7	3	8	5	6	1	2	4
5	8	2	3	1	4	6	9	7
1	4	6	7	9	2	8	5	3
4	9	7	1	3	5	2	8	6
2	5	1	6	8	7	3	4	9
6	3	8	2	4	9	7	1	5
8	1	4	9	7	3	5	6	2
7	2	5	4	6	8	9	3	1
3	6	9	5	2	1	4	7	8

Easy Solution 133

8	3	7	1	9	4	6	2	5
1	9	2	7	6	5	4	3	8
4	6	5	8	2	3	7	1	9
9	5	3	6	8	7	1	4	2
6	7	1	9	4	2	8	5	3
2	8	4	3	5	1	9	7	6
7	2	9	4	3	8	5	6	1
5	1	6	2	7	9	3	8	4
3	4	8	5	1	6	2	9	7

Easy Solution 134

3	1	9	6	7	5	2	4	8
8	6	2	4	9	3	1	7	5
4	7	5	1	8	2	9	3	6
2	8	1	5	6	7	3	9	4
9	4	3	2	1	8	5	6	7
6	5	7	3	4	9	8	1	2
1	9	6	8	5	4	7	2	3
7	3	8	9	2	6	4	5	1
5	2	4	7	3	1	6	8	9

Easy Solution 135

3	1	9	2	4	6	5	8	7
2	5	7	9	8	3	4	6	1
6	4	8	7	1	5	3	2	9
9	8	6	5	3	4	1	7	2
7	3	4	1	2	9	6	5	8
1	2	5	8	6	7	9	3	4
4	7	3	6	9	8	2	1	5
5	6	1	4	7	2	8	9	3
8	9	2	3	5	1	7	4	6

Easy Solution 136

9	7	8	2	3	5	1	6	4
6	2	3	4	7	1	9	8	5
5	4	1	9	6	8	7	2	3
7	3	4	6	1	2	8	5	9
1	9	5	3	8	7	2	4	6
2	8	6	5	9	4	3	7	1
4	1	9	8	2	6	5	3	7
8	5	7	1	4	3	6	9	2
3	6	2	7	5	9	4	1	8

Easy Solution 137

9	2	1	8	4	6	7	3	5
3	5	4	9	7	1	6	2	8
7	8	6	2	5	3	9	4	1
2	6	7	1	9	8	3	5	4
1	4	5	3	2	7	8	6	9
8	3	9	5	6	4	2	1	7
4	1	8	7	3	2	5	9	6
6	9	3	4	8	5	1	7	2
5	7	2	6	1	9	4	8	3

Easy Solution 138

8	4	9	1	7	2	3	6	5
5	1	7	3	6	9	4	8	2
2	3	6	5	8	4	9	7	1
6	9	4	8	1	7	5	2	3
7	2	5	4	3	6	8	1	9
3	8	1	2	9	5	7	4	6
4	5	3	6	2	8	1	9	7
1	7	2	9	4	3	6	5	8
9	6	8	7	5	1	2	3	4

Easy Solution 139

5	2	4	8	7	1	6	3	9
6	1	3	2	4	9	8	5	7
8	9	7	5	6	3	4	2	1
9	4	6	3	5	7	1	8	2
1	3	2	4	9	8	5	7	6
7	8	5	1	2	6	3	9	4
4	6	9	7	3	5	2	1	8
3	7	1	6	8	2	9	4	5
2	5	8	9	1	4	7	6	3

Easy Solution 140

5	7	1	3	9	4	8	2	6
9	3	6	7	8	2	4	1	5
8	4	2	5	6	1	3	9	7
4	8	3	9	1	5	6	7	2
7	6	9	2	4	8	1	5	3
1	2	5	6	7	3	9	4	8
3	1	4	8	2	7	5	6	9
6	5	7	4	3	9	2	8	1
2	9	8	1	5	6	7	3	4

Easy Solution 141

```
5 9 4 | 1 6 3 | 8 2 7
6 3 1 | 2 7 8 | 5 9 4
7 8 2 | 9 5 4 | 3 6 1
9 4 6 | 5 2 7 | 1 3 8
2 1 3 | 8 4 6 | 7 5 9
8 7 5 | 3 9 1 | 2 4 6
4 6 8 | 7 3 5 | 9 1 2
1 5 9 | 6 8 2 | 4 7 3
3 2 7 | 4 1 9 | 6 8 5
```

Easy Solution 142

```
4 9 6 | 5 1 2 | 7 8 3
3 1 2 | 8 7 6 | 4 5 9
8 7 5 | 3 9 4 | 1 2 6
1 5 9 | 6 3 7 | 2 4 8
2 3 7 | 4 8 9 | 6 1 5
6 4 8 | 1 2 5 | 3 9 7
9 8 1 | 2 6 3 | 5 7 4
7 6 4 | 9 5 1 | 8 3 2
5 2 3 | 7 4 8 | 9 6 1
```

Easy Solution 143

```
2 1 7 | 4 3 8 | 6 9 5
5 9 4 | 7 6 2 | 8 1 3
3 8 6 | 5 9 1 | 4 2 7
7 5 1 | 8 4 3 | 2 6 9
9 3 2 | 1 7 6 | 5 4 8
4 6 8 | 9 2 5 | 7 3 1
6 2 9 | 3 5 7 | 1 8 4
1 4 5 | 6 8 9 | 3 7 2
8 7 3 | 2 1 4 | 9 5 6
```

Easy Solution 144

```
7 2 6 | 1 4 3 | 8 5 9
5 1 3 | 7 8 9 | 4 6 2
8 9 4 | 2 5 6 | 1 7 3
4 3 2 | 9 6 1 | 5 8 7
9 6 8 | 3 7 5 | 2 1 4
1 7 5 | 4 2 8 | 9 3 6
2 5 7 | 6 1 4 | 3 9 8
3 4 1 | 8 9 7 | 6 2 5
6 8 9 | 5 3 2 | 7 4 1
```

Easy Solution 145

```
4 1 3 | 5 6 8 | 9 7 2
9 6 7 | 2 1 4 | 3 5 8
5 8 2 | 7 3 9 | 6 4 1
6 7 8 | 4 2 5 | 1 9 3
3 9 4 | 8 7 1 | 5 2 6
1 2 5 | 6 9 3 | 4 8 7
8 3 1 | 9 5 7 | 2 6 4
2 4 9 | 1 8 6 | 7 3 5
7 5 6 | 3 4 2 | 8 1 9
```

Easy Solution 146

```
8 6 9 | 2 3 4 | 1 7 5
7 3 1 | 5 6 9 | 4 2 8
2 4 5 | 7 1 8 | 9 3 6
3 8 6 | 9 4 2 | 7 5 1
4 9 7 | 3 5 1 | 8 6 2
1 5 2 | 6 8 7 | 3 9 4
5 7 8 | 1 9 6 | 2 4 3
9 1 3 | 4 2 5 | 6 8 7
6 2 4 | 8 7 3 | 5 1 9
```

Easy Solution 147

```
1 6 8 | 3 2 9 | 7 5 4
4 9 3 | 5 8 7 | 2 1 6
2 5 7 | 4 6 1 | 9 3 8
3 1 9 | 7 4 5 | 8 6 2
6 8 4 | 1 3 2 | 5 7 9
5 7 2 | 6 9 8 | 3 4 1
9 4 6 | 2 7 3 | 1 8 5
8 3 1 | 9 5 6 | 4 2 7
7 2 5 | 8 1 4 | 6 9 3
```

Easy Solution 148

```
1 8 3 | 6 7 9 | 2 5 4
6 5 2 | 8 4 3 | 1 9 7
7 9 4 | 5 1 2 | 8 6 3
9 2 1 | 7 6 8 | 4 3 5
4 7 8 | 9 3 5 | 6 2 1
5 3 6 | 4 2 1 | 9 7 8
8 1 9 | 2 5 7 | 3 4 6
3 4 5 | 1 9 6 | 7 8 2
2 6 7 | 3 8 4 | 5 1 9
```

Easy Solution 149

```
4 6 5 | 3 2 7 | 8 1 9
1 2 7 | 9 8 6 | 3 4 5
8 3 9 | 4 5 1 | 6 7 2
3 1 2 | 6 7 8 | 9 5 4
6 9 8 | 2 4 5 | 1 3 7
5 7 4 | 1 9 3 | 2 6 8
7 4 3 | 8 1 9 | 5 2 6
2 8 1 | 5 6 4 | 7 9 3
9 5 6 | 7 3 2 | 4 8 1
```

Easy Solution 150

```
1 6 8 | 7 4 9 | 2 5 3
3 9 2 | 1 8 5 | 4 7 6
5 4 7 | 6 3 2 | 1 9 8
9 5 1 | 2 7 3 | 6 8 4
7 2 4 | 8 5 6 | 9 3 1
8 3 6 | 9 1 4 | 5 2 7
4 1 3 | 5 2 8 | 7 6 9
6 8 5 | 4 9 7 | 3 1 2
2 7 9 | 3 6 1 | 8 4 5
```

Intermediate Solution 1

```
3 2 9 | 6 1 5 | 8 4 7
6 5 7 | 8 9 4 | 1 2 3
4 1 8 | 2 7 3 | 9 6 5
2 7 4 | 3 8 6 | 5 9 1
1 9 6 | 7 5 2 | 4 3 8
5 8 3 | 1 4 9 | 2 7 6
8 3 2 | 4 6 1 | 7 5 9
9 6 1 | 5 2 7 | 3 8 4
7 4 5 | 9 3 8 | 6 1 2
```

Intermediate Solution 2

```
8 1 9 | 5 6 3 | 2 4 7
5 2 6 | 4 8 7 | 9 3 1
4 7 3 | 2 1 9 | 5 6 8
6 9 7 | 1 5 8 | 3 2 4
3 4 2 | 9 7 6 | 1 8 5
1 8 5 | 3 4 2 | 7 9 6
2 5 1 | 8 9 4 | 6 7 3
9 6 8 | 7 3 1 | 4 5 2
7 3 4 | 6 2 5 | 8 1 9
```

Intermediate Solution 3

```
9 7 5 | 4 3 2 | 8 6 1
2 6 8 | 9 5 1 | 3 4 7
3 1 4 | 6 7 8 | 2 9 5
7 3 1 | 5 8 6 | 9 2 4
8 4 9 | 7 2 3 | 5 1 6
6 5 2 | 1 4 9 | 7 3 8
4 9 3 | 8 1 5 | 6 7 2
5 2 7 | 3 6 4 | 1 8 9
1 8 6 | 2 9 7 | 4 5 3
```

Intermediate Solution 4

```
7 4 9 | 5 6 1 | 3 8 2
5 3 2 | 8 9 7 | 1 6 4
6 8 1 | 4 3 2 | 5 9 7
3 7 6 | 1 4 9 | 2 5 8
9 2 5 | 3 8 6 | 7 4 1
8 1 4 | 7 2 5 | 9 3 6
2 9 3 | 6 7 4 | 8 1 5
1 6 7 | 9 5 8 | 4 2 3
4 5 8 | 2 1 3 | 6 7 9
```

Intermediate Solution 5

```
9 6 5 | 1 4 8 | 2 7 3
4 8 7 | 6 2 3 | 1 5 9
2 1 3 | 7 5 9 | 4 8 6
7 5 9 | 3 6 4 | 8 2 1
8 4 1 | 9 7 2 | 3 6 5
3 2 6 | 8 1 5 | 9 4 7
5 9 8 | 2 3 6 | 7 1 4
6 7 2 | 4 9 1 | 5 3 8
1 3 4 | 5 8 7 | 6 9 2
```

Intermediate Solution 6

```
4 9 1 | 3 7 2 | 5 6 8
6 2 7 | 8 4 5 | 1 3 9
5 3 8 | 6 9 1 | 4 7 2
2 1 4 | 7 8 6 | 9 5 3
8 5 3 | 4 2 9 | 6 1 7
9 7 6 | 1 5 3 | 8 2 4
7 8 2 | 5 6 4 | 3 9 1
3 4 5 | 9 1 7 | 2 8 6
1 6 9 | 2 3 8 | 7 4 5
```

Intermediate Solution 7

```
6 8 9 | 4 3 1 | 7 2 5
3 1 7 | 2 5 8 | 6 4 9
2 4 5 | 9 7 6 | 3 8 1
1 5 6 | 3 8 2 | 9 7 4
8 9 2 | 7 6 4 | 5 1 3
4 7 3 | 5 1 9 | 8 6 2
9 3 4 | 8 2 7 | 1 5 6
5 6 8 | 1 4 3 | 2 9 7
7 2 1 | 6 9 5 | 4 3 8
```

Intermediate Solution 8

```
9 1 2 | 6 3 5 | 4 8 7
8 6 5 | 4 7 2 | 1 9 3
3 4 7 | 9 8 1 | 5 2 6
6 3 9 | 5 2 8 | 7 4 1
4 2 1 | 7 6 9 | 8 3 5
7 5 8 | 1 4 3 | 9 6 2
2 9 3 | 8 1 7 | 6 5 4
5 7 6 | 2 9 4 | 3 1 8
1 8 4 | 3 5 6 | 2 7 9
```

Intermediate Solution 9

```
9 1 3 | 2 5 4 | 8 7 6
5 7 2 | 8 1 6 | 9 3 4
4 6 8 | 7 9 3 | 2 5 1
2 8 6 | 1 3 7 | 4 9 5
1 4 5 | 9 8 2 | 3 6 7
3 9 7 | 4 6 5 | 1 8 2
7 3 9 | 5 2 1 | 6 4 8
8 5 1 | 6 4 9 | 7 2 3
6 2 4 | 3 7 8 | 5 1 9
```

Intermediate Solution 10

```
5 3 7 | 4 6 9 | 2 8 1
1 4 2 | 8 7 3 | 6 9 5
8 9 6 | 5 2 1 | 3 4 7
6 7 9 | 2 4 5 | 1 3 8
4 5 1 | 3 8 7 | 9 2 6
3 2 8 | 9 1 6 | 5 7 4
9 1 3 | 7 5 8 | 4 6 2
7 6 4 | 1 3 2 | 8 5 9
2 8 5 | 6 9 4 | 7 1 3
```

Intermediate Solution 11

2	1	8	9	5	3	6	7	4
7	5	9	8	4	6	1	2	3
3	6	4	7	1	2	9	5	8
6	7	5	4	2	9	3	8	1
8	3	1	6	7	5	4	9	2
4	9	2	3	8	1	7	6	5
1	2	6	5	3	7	8	4	9
9	8	3	2	6	4	5	1	7
5	4	7	1	9	8	2	3	6

Intermediate Solution 12

9	1	8	7	6	3	5	2	4
4	2	7	5	1	9	6	8	3
6	3	5	2	4	8	9	7	1
3	6	2	9	5	7	1	4	8
8	9	1	6	2	4	7	3	5
5	7	4	3	8	1	2	9	6
7	5	9	8	3	6	4	1	2
2	4	3	1	9	5	8	6	7
1	8	6	4	7	2	3	5	9

Intermediate Solution 13

4	8	5	3	7	6	2	9	1
6	1	3	9	8	2	4	7	5
7	2	9	1	5	4	3	6	8
3	7	8	6	9	1	5	4	2
9	6	4	2	3	5	1	8	7
1	5	2	7	4	8	9	3	6
5	9	1	4	6	7	8	2	3
8	4	6	5	2	3	7	1	9
2	3	7	8	1	9	6	5	4

Intermediate Solution 14

7	2	1	4	5	9	6	8	3
6	5	8	3	7	2	1	4	9
9	4	3	1	8	6	5	7	2
4	1	9	6	3	8	7	2	5
2	3	6	7	1	5	8	9	4
8	7	5	9	2	4	3	6	1
3	9	4	5	6	7	2	1	8
5	8	7	2	9	1	4	3	6
1	6	2	8	4	3	9	5	7

Intermediate Solution 15

3	8	6	4	5	2	7	9	1
7	1	2	9	6	3	5	4	8
4	9	5	8	7	1	6	3	2
8	3	9	1	4	7	2	6	5
1	5	4	2	3	6	9	8	7
2	6	7	5	9	8	3	1	4
5	2	3	6	1	4	8	7	9
6	4	8	7	2	9	1	5	3
9	7	1	3	8	5	4	2	6

Intermediate Solution 16

4	7	9	5	3	8	2	6	1
2	1	3	7	6	4	8	9	5
8	6	5	9	1	2	3	4	7
7	3	2	4	8	6	1	5	9
5	4	1	2	9	7	6	8	3
6	9	8	3	5	1	7	2	4
1	8	7	6	4	9	5	3	2
3	2	4	8	7	5	9	1	6
9	5	6	1	2	3	4	7	8

Intermediate Solution 17

9	7	4	3	1	2	6	5	8
3	8	1	7	5	6	9	4	2
6	5	2	9	4	8	7	3	1
5	2	3	4	6	7	1	8	9
8	4	6	2	9	1	3	7	5
1	9	7	8	3	5	4	2	6
4	6	5	1	8	3	2	9	7
2	3	8	6	7	9	5	1	4
7	1	9	5	2	4	8	6	3

Intermediate Solution 18

9	3	2	5	8	4	6	7	1
4	6	1	3	2	7	8	9	5
7	5	8	6	9	1	4	3	2
5	4	7	8	3	2	1	6	9
6	8	3	9	1	5	7	2	4
1	2	9	7	4	6	5	8	3
3	9	5	4	7	8	2	1	6
2	7	6	1	5	3	9	4	8
8	1	4	2	6	9	3	5	7

Intermediate Solution 19

8	3	9	4	5	6	1	7	2
6	4	7	2	1	3	5	9	8
2	1	5	8	7	9	3	4	6
4	7	2	1	8	5	6	3	9
5	8	6	3	9	4	2	1	7
3	9	1	6	2	7	8	5	4
1	6	4	7	3	2	9	8	5
9	2	8	5	4	1	7	6	3
7	5	3	9	6	8	4	2	1

Intermediate Solution 20

9	7	4	5	8	3	2	6	1
3	1	8	6	2	4	7	5	9
5	2	6	7	1	9	4	8	3
6	5	3	9	7	8	1	2	4
7	9	2	1	4	5	8	3	6
8	4	1	2	3	6	9	7	5
1	8	5	3	9	7	6	4	2
2	3	7	4	6	1	5	9	8
4	6	9	8	5	2	3	1	7

Intermediate Solution 21

2	9	6	7	3	1	5	4	8
8	1	3	2	4	5	6	7	9
4	7	5	6	9	8	1	3	2
5	3	1	4	7	9	2	8	6
9	6	8	1	2	3	7	5	4
7	2	4	5	8	6	3	9	1
1	4	9	3	5	2	8	6	7
6	5	7	8	1	4	9	2	3
3	8	2	9	6	7	4	1	5

Intermediate Solution 22

2	7	8	4	1	9	6	5	3
4	9	3	5	6	7	2	8	1
1	5	6	8	3	2	4	9	7
9	6	1	7	5	4	3	2	8
8	2	5	3	9	1	7	6	4
7	3	4	6	2	8	5	1	9
3	1	9	2	4	5	8	7	6
6	8	2	9	7	3	1	4	5
5	4	7	1	8	6	9	3	2

Intermediate Solution 23

9	7	4	2	1	3	6	8	5
6	1	2	8	5	9	7	3	4
5	8	3	7	6	4	2	1	9
3	9	1	6	4	8	5	2	7
7	2	6	5	9	1	3	4	8
8	4	5	3	2	7	9	6	1
4	3	9	1	7	2	8	5	6
1	6	8	9	3	5	4	7	2
2	5	7	4	8	6	1	9	3

Intermediate Solution 24

8	1	4	9	2	5	3	7	6
2	6	7	3	4	1	8	5	9
3	9	5	8	7	6	4	2	1
5	8	1	4	9	2	6	3	7
6	7	9	5	8	3	1	4	2
4	2	3	1	6	7	9	8	5
7	3	2	6	1	4	5	9	8
9	4	6	7	5	8	2	1	3
1	5	8	2	3	9	7	6	4

Intermediate Solution 25

2	6	4	3	9	1	7	5	8
9	1	7	5	4	8	2	6	3
8	3	5	7	6	2	9	1	4
1	9	8	2	3	7	6	4	5
5	2	6	1	8	4	3	9	7
7	4	3	6	5	9	8	2	1
3	7	2	4	1	6	5	8	9
4	5	9	8	2	3	1	7	6
6	8	1	9	7	5	4	3	2

Intermediate Solution 26

6	7	8	2	9	4	5	1	3
4	2	3	8	5	1	7	9	6
9	1	5	6	3	7	4	2	8
7	5	6	3	8	9	1	4	2
3	9	1	4	7	2	6	8	5
8	4	2	1	6	5	9	3	7
2	6	9	5	4	8	3	7	1
1	3	7	9	2	6	8	5	4
5	8	4	7	1	3	2	6	9

Intermediate Solution 27

8	1	4	7	3	6	9	5	2
9	6	3	2	4	5	8	1	7
7	2	5	1	9	8	6	3	4
1	3	8	5	7	2	4	6	9
2	9	6	4	1	3	5	7	8
4	5	7	6	8	9	3	2	1
3	4	2	9	6	7	1	8	5
5	8	1	3	2	4	7	9	6
6	7	9	8	5	1	2	4	3

Intermediate Solution 28

8	3	6	9	5	1	7	2	4
5	1	4	2	8	7	9	6	3
7	2	9	4	3	6	1	8	5
2	4	8	3	1	5	6	7	9
3	9	7	6	4	8	2	5	1
6	5	1	7	9	2	4	3	8
4	6	5	8	2	9	3	1	7
9	8	2	1	7	3	5	4	6
1	7	3	5	6	4	8	9	2

Intermediate Solution 29

5	3	9	1	6	8	7	2	4
1	7	6	4	9	2	8	3	5
8	4	2	7	5	3	6	1	9
7	9	3	6	1	5	2	4	8
4	8	1	3	2	7	9	5	6
2	6	5	8	4	9	3	7	1
9	2	4	5	3	6	1	8	7
6	1	7	2	8	4	5	9	3
3	5	8	9	7	1	4	6	2

Intermediate Solution 30

9	2	4	8	5	3	6	7	1
7	5	3	1	6	4	8	9	2
6	8	1	7	9	2	3	5	4
3	4	8	9	1	7	5	2	6
1	6	5	2	3	8	7	4	9
2	9	7	6	4	5	1	3	8
8	3	6	5	2	9	4	1	7
4	7	9	3	8	1	2	6	5
5	1	2	4	7	6	9	8	3

Intermediate Solution 31

7	9	8	1	5	2	6	3	4
2	4	5	6	9	3	7	1	8
1	3	6	8	7	4	2	9	5
5	7	4	9	6	1	8	2	3
3	6	2	4	8	5	1	7	9
8	1	9	2	3	7	5	4	6
4	5	7	3	2	8	9	6	1
6	8	1	7	4	9	3	5	2
9	2	3	5	1	6	4	8	7

Intermediate Solution 32

2	9	7	5	1	3	4	8	6
3	5	8	4	2	6	7	1	9
6	4	1	9	7	8	2	5	3
5	8	9	7	3	2	6	4	1
7	2	3	1	6	4	5	9	8
1	6	4	8	9	5	3	2	7
4	1	2	3	8	7	9	6	5
9	3	6	2	5	1	8	7	4
8	7	5	6	4	9	1	3	2

Intermediate Solution 33

5	7	8	1	9	4	3	6	2
6	9	4	2	5	3	1	8	7
3	1	2	7	6	8	5	4	9
9	8	1	6	7	5	2	3	4
4	2	3	9	8	1	6	7	5
7	6	5	3	4	2	8	9	1
8	4	7	5	2	6	9	1	3
2	3	6	4	1	9	7	5	8
1	5	9	8	3	7	4	2	6

Intermediate Solution 34

1	3	2	5	8	4	7	9	6
6	4	7	2	9	1	5	3	8
8	5	9	7	3	6	2	1	4
2	7	5	3	4	9	6	8	1
3	1	6	8	7	5	9	4	2
9	8	4	6	1	2	3	7	5
5	9	1	4	6	7	8	2	3
4	2	8	9	5	3	1	6	7
7	6	3	1	2	8	4	5	9

Intermediate Solution 35

7	5	4	8	3	6	1	2	9
6	2	3	9	7	1	5	8	4
8	1	9	2	5	4	7	3	6
2	8	1	4	9	5	6	7	3
4	6	5	7	8	3	2	9	1
9	3	7	1	6	2	8	4	5
5	9	6	3	2	7	4	1	8
1	7	8	6	4	9	3	5	2
3	4	2	5	1	8	9	6	7

Intermediate Solution 36

5	3	6	2	4	9	1	7	8
8	4	1	7	6	3	2	5	9
2	7	9	8	5	1	6	3	4
3	5	4	9	8	6	7	1	2
9	8	2	5	1	7	3	4	6
6	1	7	3	2	4	8	9	5
4	6	8	1	7	5	9	2	3
7	9	5	6	3	2	4	8	1
1	2	3	4	9	8	5	6	7

Intermediate Solution 37

7	2	9	6	8	1	4	5	3
5	8	4	2	3	9	1	6	7
1	3	6	5	7	4	9	2	8
9	7	2	8	4	6	5	3	1
3	5	1	9	2	7	6	8	4
6	4	8	1	5	3	2	7	9
4	6	7	3	1	2	8	9	5
2	1	5	7	9	8	3	4	6
8	9	3	4	6	5	7	1	2

Intermediate Solution 38

9	6	3	5	2	4	8	7	1
7	1	4	3	8	6	5	2	9
2	8	5	9	1	7	3	4	6
6	7	2	8	4	9	1	3	5
3	9	8	7	5	1	2	6	4
5	4	1	6	3	2	9	8	7
4	3	7	1	9	8	6	5	2
1	5	6	2	7	3	4	9	8
8	2	9	4	6	5	7	1	3

Intermediate Solution 39

4	1	2	3	7	6	5	8	9
9	5	6	8	2	1	3	7	4
3	8	7	9	5	4	1	6	2
7	3	9	2	8	5	4	1	6
8	4	1	6	3	7	9	2	5
6	2	5	4	1	9	8	3	7
1	6	3	5	9	2	7	4	8
5	7	4	1	6	8	2	9	3
2	9	8	7	4	3	6	5	1

Intermediate Solution 40

6	4	7	5	9	1	3	2	8
1	8	5	3	2	7	4	9	6
3	9	2	8	4	6	5	1	7
2	7	8	6	5	4	1	3	9
5	6	4	1	3	9	7	8	2
9	1	3	7	8	2	6	4	5
8	3	1	2	7	5	9	6	4
7	2	9	4	6	3	8	5	1
4	5	6	9	1	8	2	7	3

Intermediate Solution 41

5	3	8	7	9	4	2	1	6
2	6	1	5	8	3	4	7	9
4	9	7	1	6	2	3	5	8
3	1	5	2	4	9	8	6	7
9	7	6	8	3	1	5	2	4
8	2	4	6	5	7	1	9	3
6	4	3	9	2	5	7	8	1
7	5	9	4	1	8	6	3	2
1	8	2	3	7	6	9	4	5

Intermediate Solution 42

5	1	9	8	4	2	7	6	3
7	8	6	9	3	1	4	5	2
4	3	2	6	7	5	1	8	9
1	9	8	5	2	4	3	7	6
3	4	5	7	1	6	2	9	8
2	6	7	3	9	8	5	1	4
6	2	4	1	5	9	8	3	7
8	7	1	2	6	3	9	4	5
9	5	3	4	8	7	6	2	1

Intermediate Solution 43

9	8	5	7	3	4	2	1	6
7	3	1	6	2	9	8	5	4
6	4	2	1	5	8	3	9	7
4	9	7	3	8	1	6	2	5
2	1	8	5	6	7	4	3	9
5	6	3	9	4	2	7	8	1
8	7	9	2	1	6	5	4	3
1	5	4	8	7	3	9	6	2
3	2	6	4	9	5	1	7	8

Intermediate Solution 44

8	9	7	5	2	1	4	3	6
3	1	4	8	9	6	5	2	7
5	2	6	4	7	3	8	1	9
9	6	3	1	5	2	7	8	4
7	8	2	6	4	9	1	5	3
1	4	5	3	8	7	9	6	2
2	5	8	9	6	4	3	7	1
4	7	1	2	3	5	6	9	8
6	3	9	7	1	8	2	4	5

Intermediate Solution 45

9	2	5	8	4	1	3	6	7
6	1	3	9	2	7	4	5	8
4	8	7	5	6	3	1	9	2
8	5	2	3	1	6	9	7	4
7	9	1	2	5	4	8	3	6
3	4	6	7	9	8	5	2	1
1	3	4	6	7	5	2	8	9
5	6	9	1	8	2	7	4	3
2	7	8	4	3	9	6	1	5

Intermediate Solution 46

2	5	8	4	9	1	7	3	6
9	1	3	7	5	6	4	8	2
6	4	7	3	8	2	1	9	5
7	3	1	5	2	9	6	4	8
4	9	2	6	3	8	5	1	7
5	8	6	1	7	4	9	2	3
3	2	4	9	6	5	8	7	1
8	6	9	2	1	7	3	5	4
1	7	5	8	4	3	2	6	9

Intermediate Solution 47

2	6	5	3	8	7	9	4	1
9	3	7	4	6	1	2	8	5
8	1	4	9	2	5	3	7	6
5	9	3	1	4	6	8	2	7
6	7	8	5	3	2	4	1	9
4	2	1	8	7	9	5	6	3
1	8	9	6	5	4	7	3	2
7	4	6	2	9	3	1	5	8
3	5	2	7	1	8	6	9	4

Intermediate Solution 48

4	9	8	3	7	6	5	1	2
7	5	1	2	8	4	6	3	9
6	3	2	9	5	1	7	8	4
9	2	7	5	6	3	8	4	1
3	4	5	1	2	8	9	6	7
1	8	6	4	9	7	3	2	5
5	7	3	8	4	2	1	9	6
8	6	4	7	1	9	2	5	3
2	1	9	6	3	5	4	7	8

Intermediate Solution 49

6	8	2	4	5	9	7	1	3
1	3	4	8	7	6	9	2	5
9	7	5	1	2	3	6	4	8
2	1	6	7	9	8	5	3	4
5	4	8	3	6	2	1	9	7
3	9	7	5	1	4	8	6	2
7	2	1	6	3	5	4	8	9
4	5	9	2	8	1	3	7	6
8	6	3	9	4	7	2	5	1

Intermediate Solution 50

8	1	9	2	5	7	6	4	3
4	7	2	1	6	3	5	8	9
3	6	5	4	9	8	1	7	2
7	9	1	3	2	4	8	5	6
5	3	6	8	7	9	4	2	1
2	4	8	5	1	6	3	9	7
6	5	3	7	8	2	9	1	4
1	2	4	9	3	5	7	6	8
9	8	7	6	4	1	2	3	5

Intermediate Solution 51

5	3	7	2	6	8	1	9	4
2	6	9	7	4	1	5	3	8
1	4	8	9	3	5	6	2	7
8	7	4	3	5	9	2	6	1
9	1	3	8	2	6	7	4	5
6	5	2	4	1	7	3	8	9
7	2	1	6	8	4	9	5	3
4	9	6	5	7	3	8	1	2
3	8	5	1	9	2	4	7	6

Intermediate Solution 52

7	6	4	3	5	2	9	8	1
9	5	8	4	1	7	3	6	2
1	3	2	6	8	9	5	7	4
4	1	3	9	7	5	6	2	8
5	7	9	8	2	6	1	4	3
2	8	6	1	4	3	7	9	5
8	9	5	2	6	1	4	3	7
3	4	1	7	9	8	2	5	6
6	2	7	5	3	4	8	1	9

Intermediate Solution 53

3	6	1	5	9	8	2	4	7
9	4	5	7	2	3	8	6	1
7	2	8	1	4	6	3	5	9
4	5	6	8	7	1	9	3	2
2	1	3	4	6	9	5	7	8
8	7	9	2	3	5	4	1	6
5	3	7	6	8	2	1	9	4
6	9	2	3	1	4	7	8	5
1	8	4	9	5	7	6	2	3

Intermediate Solution 54

9	1	5	6	4	3	2	7	8
2	8	4	5	9	7	6	3	1
7	3	6	2	8	1	9	4	5
4	2	3	8	1	5	7	9	6
8	5	9	7	6	4	3	1	2
1	6	7	3	2	9	5	8	4
6	4	1	9	3	2	8	5	7
3	7	2	1	5	8	4	6	9
5	9	8	4	7	6	1	2	3

Intermediate Solution 55

7	9	3	1	4	6	2	8	5
5	2	1	8	3	7	9	4	6
4	6	8	2	5	9	3	7	1
8	1	4	6	9	3	7	5	2
3	5	6	4	7	2	1	9	8
2	7	9	5	8	1	6	3	4
1	3	5	9	6	8	4	2	7
6	8	7	3	2	4	5	1	9
9	4	2	7	1	5	8	6	3

Intermediate Solution 56

1	5	3	7	2	4	6	9	8
9	7	6	1	8	5	2	4	3
8	2	4	3	9	6	1	5	7
5	1	7	8	6	3	4	2	9
6	4	2	5	7	9	3	8	1
3	9	8	2	4	1	5	7	6
7	6	5	4	1	8	9	3	2
2	3	9	6	5	7	8	1	4
4	8	1	9	3	2	7	6	5

Intermediate Solution 57

8	6	7	1	5	9	4	3	2
3	2	9	7	4	8	6	1	5
1	5	4	2	3	6	8	9	7
6	7	2	4	8	3	1	5	9
5	1	8	9	6	7	2	4	3
9	4	3	5	2	1	7	6	8
7	3	6	8	1	5	9	2	4
4	8	1	3	9	2	5	7	6
2	9	5	6	7	4	3	8	1

Intermediate Solution 58

4	2	1	9	8	3	7	5	6
3	7	5	6	4	1	9	8	2
6	8	9	7	2	5	4	1	3
1	4	3	2	6	8	5	7	9
8	5	7	3	1	9	2	6	4
9	6	2	4	5	7	8	3	1
2	9	8	1	7	6	3	4	5
7	1	4	5	3	2	6	9	8
5	3	6	8	9	4	1	2	7

Intermediate Solution 59

5	3	7	1	8	4	6	9	2
2	8	6	3	9	5	1	4	7
1	4	9	6	2	7	8	3	5
9	1	4	5	3	2	7	8	6
7	2	3	8	1	6	9	5	4
8	6	5	7	4	9	3	2	1
6	9	8	2	5	1	4	7	3
4	7	2	9	6	3	5	1	8
3	5	1	4	7	8	2	6	9

Intermediate Solution 60

1	5	6	9	3	8	2	4	7
3	2	7	6	4	5	1	8	9
4	9	8	2	7	1	5	3	6
9	7	4	8	2	3	6	1	5
6	1	2	4	5	7	3	9	8
8	3	5	1	6	9	4	7	2
2	6	1	7	9	4	8	5	3
7	4	3	5	8	6	9	2	1
5	8	9	3	1	2	7	6	4

Intermediate Solution 61

8	3	4	9	7	6	5	2	1
1	9	6	4	2	5	3	7	8
7	2	5	1	3	8	6	9	4
9	6	8	3	4	2	1	5	7
3	1	2	6	5	7	4	8	9
4	5	7	8	1	9	2	3	6
5	8	1	7	6	3	9	4	2
6	7	3	2	9	4	8	1	5
2	4	9	5	8	1	7	6	3

Intermediate Solution 62

6	3	2	9	1	8	5	7	4
4	5	7	6	2	3	8	9	1
9	1	8	7	5	4	2	3	6
2	6	4	5	7	9	1	8	3
3	9	5	1	8	2	4	6	7
7	8	1	3	4	6	9	5	2
5	2	9	4	3	7	6	1	8
1	4	3	8	6	5	7	2	9
8	7	6	2	9	1	3	4	5

Intermediate Solution 63

5	2	8	9	6	7	1	3	4
9	7	6	4	1	3	8	5	2
4	3	1	2	8	5	7	9	6
2	1	4	3	5	6	9	7	8
7	9	3	8	2	4	5	6	1
6	8	5	7	9	1	2	4	3
8	5	2	6	4	9	3	1	7
3	6	9	1	7	8	4	2	5
1	4	7	5	3	2	6	8	9

Intermediate Solution 64

3	6	1	8	5	7	4	9	2
8	7	5	2	4	9	1	6	3
4	2	9	3	6	1	7	5	8
1	5	4	9	3	2	6	8	7
2	8	3	6	7	4	5	1	9
7	9	6	5	1	8	3	2	4
9	4	7	1	2	5	8	3	6
6	1	8	7	9	3	2	4	5
5	3	2	4	8	6	9	7	1

Intermediate Solution 65

2	5	9	1	7	8	3	4	6
3	8	1	4	2	6	5	7	9
7	4	6	9	3	5	2	8	1
4	6	7	5	8	1	9	3	2
9	3	5	2	6	7	4	1	8
8	1	2	3	9	4	7	6	5
5	7	4	6	1	2	8	9	3
6	2	3	8	4	9	1	5	7
1	9	8	7	5	3	6	2	4

Intermediate Solution 66

6	1	7	9	3	4	8	5	2
2	3	4	5	1	8	9	6	7
8	9	5	7	2	6	4	3	1
7	8	1	2	4	5	6	9	3
5	2	6	1	9	3	7	8	4
3	4	9	6	8	7	2	1	5
1	7	8	3	6	2	5	4	9
9	6	2	4	5	1	3	7	8
4	5	3	8	7	9	1	2	6

Intermediate Solution 67

1	6	7	2	9	5	8	4	3
4	2	5	3	6	8	9	1	7
8	9	3	7	1	4	6	2	5
3	7	1	9	4	6	5	8	2
6	4	9	8	5	2	3	7	1
5	8	2	1	7	3	4	9	6
2	1	4	5	3	9	7	6	8
7	5	6	4	8	1	2	3	9
9	3	8	6	2	7	1	5	4

Intermediate Solution 68

5	2	1	4	3	9	8	6	7
8	7	6	2	5	1	4	9	3
9	3	4	6	8	7	5	2	1
2	9	5	7	1	6	3	4	8
6	4	3	8	2	5	1	7	9
1	8	7	3	9	4	2	5	6
3	6	8	9	4	2	7	1	5
7	1	2	5	6	3	9	8	4
4	5	9	1	7	8	6	3	2

Intermediate Solution 69

5	7	2	9	4	8	3	6	1
1	8	3	7	6	5	9	4	2
9	4	6	2	3	1	7	5	8
2	1	9	5	7	6	8	3	4
4	3	7	8	1	2	5	9	6
8	6	5	3	9	4	1	2	7
3	2	4	1	8	9	6	7	5
7	5	1	6	2	3	4	8	9
6	9	8	4	5	7	2	1	3

Intermediate Solution 70

7	5	8	3	9	1	2	4	6
4	1	6	7	2	5	9	8	3
9	3	2	4	8	6	1	7	5
1	9	5	6	3	4	7	2	8
2	6	3	1	7	8	4	5	9
8	7	4	9	5	2	6	3	1
3	2	7	5	1	9	8	6	4
5	4	1	8	6	7	3	9	2
6	8	9	2	4	3	5	1	7

Intermediate Solution 71

```
6 1 5 | 8 7 3 | 4 9 2
4 7 2 | 6 1 9 | 5 3 8
9 8 3 | 4 5 2 | 7 1 6
------+-------+------
5 9 1 | 7 2 6 | 8 4 3
7 4 8 | 9 3 1 | 2 6 5
2 3 6 | 5 4 8 | 9 7 1
------+-------+------
1 2 4 | 3 8 7 | 6 5 9
8 5 9 | 1 6 4 | 3 2 7
3 6 7 | 2 9 5 | 1 8 4
```

Intermediate Solution 72

```
5 6 3 | 4 9 2 | 8 7 1
4 2 7 | 1 8 3 | 6 5 9
8 9 1 | 6 5 7 | 2 3 4
------+-------+------
7 4 6 | 3 1 5 | 9 2 8
2 1 8 | 7 4 9 | 5 6 3
9 3 5 | 2 6 8 | 4 1 7
------+-------+------
1 8 9 | 5 3 6 | 7 4 2
3 5 2 | 9 7 4 | 1 8 6
6 7 4 | 8 2 1 | 3 9 5
```

Intermediate Solution 73

```
7 8 2 | 4 5 9 | 3 6 1
6 4 5 | 3 1 7 | 9 2 8
9 3 1 | 8 6 2 | 5 4 7
------+-------+------
5 6 9 | 2 3 1 | 7 8 4
4 2 3 | 5 7 8 | 1 9 6
8 1 7 | 6 9 4 | 2 5 3
------+-------+------
1 7 4 | 9 2 6 | 8 3 5
3 9 6 | 1 8 5 | 4 7 2
2 5 8 | 7 4 3 | 6 1 9
```

Intermediate Solution 74

```
3 1 6 | 8 2 5 | 7 9 4
8 2 7 | 9 3 4 | 1 6 5
5 9 4 | 1 7 6 | 2 3 8
------+-------+------
1 6 2 | 3 5 7 | 4 8 9
9 7 3 | 4 8 1 | 6 5 2
4 5 8 | 2 6 9 | 3 7 1
------+-------+------
7 8 9 | 6 4 2 | 5 1 3
6 4 1 | 5 9 3 | 8 2 7
2 3 5 | 7 1 8 | 9 4 6
```

Intermediate Solution 75

```
2 7 6 | 3 4 5 | 8 1 9
9 8 1 | 7 6 2 | 5 3 4
5 3 4 | 8 9 1 | 2 6 7
------+-------+------
8 2 9 | 5 1 7 | 6 4 3
4 5 3 | 6 2 9 | 1 7 8
6 1 7 | 4 3 8 | 9 5 2
------+-------+------
3 9 5 | 2 7 6 | 4 8 1
7 6 2 | 1 8 4 | 3 9 5
1 4 8 | 9 5 3 | 7 2 6
```

Intermediate Solution 76

```
5 1 9 | 8 4 7 | 2 3 6
3 8 2 | 6 9 5 | 4 1 7
4 7 6 | 2 3 1 | 9 5 8
------+-------+------
7 4 8 | 9 1 2 | 5 6 3
2 6 3 | 5 8 4 | 7 9 1
1 9 5 | 7 6 3 | 8 4 2
------+-------+------
9 3 1 | 4 7 8 | 6 2 5
6 5 7 | 1 2 9 | 3 8 4
8 2 4 | 3 5 6 | 1 7 9
```

Intermediate Solution 77

```
1 3 4 | 7 2 5 | 9 8 6
2 7 8 | 9 6 1 | 5 3 4
5 9 6 | 3 8 4 | 2 1 7
------+-------+------
4 8 9 | 2 5 6 | 3 7 1
7 1 5 | 4 3 8 | 6 9 2
3 6 2 | 1 9 7 | 8 4 5
------+-------+------
8 2 7 | 5 1 3 | 4 6 9
6 5 1 | 8 4 9 | 7 2 3
9 4 3 | 6 7 2 | 1 5 8
```

Intermediate Solution 78

```
5 6 2 | 3 7 9 | 8 4 1
3 8 7 | 1 4 5 | 9 6 2
4 1 9 | 6 8 2 | 7 3 5
------+-------+------
8 2 5 | 9 6 7 | 4 1 3
6 3 4 | 5 1 8 | 2 9 7
7 9 1 | 4 2 3 | 6 5 8
------+-------+------
2 5 3 | 7 9 6 | 1 8 4
1 7 6 | 8 3 4 | 5 2 9
9 4 8 | 2 5 1 | 3 7 6
```

Intermediate Solution 79

```
4 5 2 | 1 8 6 | 9 3 7
1 6 8 | 7 3 9 | 2 5 4
9 3 7 | 4 2 5 | 8 1 6
------+-------+------
5 9 3 | 8 6 7 | 1 4 2
8 1 6 | 9 4 2 | 3 7 5
2 7 4 | 5 1 3 | 6 9 8
------+-------+------
6 4 9 | 2 7 1 | 5 8 3
7 2 1 | 3 5 8 | 4 6 9
3 8 5 | 6 9 4 | 7 2 1
```

Intermediate Solution 80

```
8 6 2 | 1 3 5 | 7 4 9
1 4 3 | 8 7 9 | 6 2 5
7 5 9 | 2 4 6 | 3 1 8
------+-------+------
2 7 6 | 3 9 1 | 5 8 4
5 9 4 | 7 6 8 | 1 3 2
3 8 1 | 4 5 2 | 9 6 7
------+-------+------
4 1 5 | 9 2 3 | 8 7 6
6 2 8 | 5 1 7 | 4 9 3
9 3 7 | 6 8 4 | 2 5 1
```

Intermediate Solution 81

```
8 5 9 | 1 3 4 | 2 7 6
3 4 6 | 7 2 5 | 9 8 1
2 1 7 | 9 6 8 | 4 3 5
------+-------+------
7 9 5 | 8 4 6 | 3 1 2
4 2 3 | 5 9 1 | 8 6 7
1 6 8 | 2 7 3 | 5 4 9
------+-------+------
5 7 4 | 3 1 9 | 6 2 8
6 8 2 | 4 5 7 | 1 9 3
9 3 1 | 6 8 2 | 7 5 4
```

Intermediate Solution 82

```
7 8 5 | 1 4 9 | 2 3 6
1 6 3 | 2 8 7 | 9 4 5
2 9 4 | 3 5 6 | 8 1 7
------+-------+------
9 1 6 | 7 3 2 | 5 8 4
8 5 7 | 4 6 1 | 3 9 2
4 3 2 | 8 9 5 | 7 6 1
------+-------+------
3 7 8 | 6 2 4 | 1 5 9
5 4 1 | 9 7 3 | 6 2 8
6 2 9 | 5 1 8 | 4 7 3
```

Intermediate Solution 83

```
7 3 6 | 8 2 4 | 9 5 1
4 1 2 | 5 9 7 | 8 3 6
5 8 9 | 3 6 1 | 2 4 7
------+-------+------
8 6 5 | 9 4 3 | 1 7 2
3 2 7 | 6 1 5 | 4 8 9
9 4 1 | 2 7 8 | 5 6 3
------+-------+------
6 5 8 | 1 3 9 | 7 2 4
1 7 3 | 4 8 2 | 6 9 5
2 9 4 | 7 5 6 | 3 1 8
```

Intermediate Solution 84

```
6 1 5 | 3 8 7 | 4 9 2
3 4 9 | 6 2 1 | 5 7 8
8 2 7 | 9 5 4 | 1 6 3
------+-------+------
2 6 1 | 4 7 9 | 3 8 5
7 3 4 | 5 6 8 | 9 2 1
5 9 8 | 2 1 3 | 6 4 7
------+-------+------
1 7 6 | 8 4 5 | 2 3 9
9 8 2 | 1 3 6 | 7 5 4
4 5 3 | 7 9 2 | 8 1 6
```

Intermediate Solution 85

```
8 1 4 | 3 2 6 | 5 9 7
6 9 2 | 4 5 7 | 8 3 1
7 3 5 | 9 8 1 | 6 4 2
------+-------+------
9 8 7 | 5 3 4 | 1 2 6
1 5 3 | 2 6 8 | 4 7 9
2 4 6 | 7 1 9 | 3 5 8
------+-------+------
5 7 1 | 8 9 3 | 2 6 4
3 6 9 | 1 4 2 | 7 8 5
4 2 8 | 6 7 5 | 9 1 3
```

Intermediate Solution 86

```
9 6 5 | 2 4 8 | 3 7 1
8 1 2 | 3 7 5 | 4 6 9
4 7 3 | 9 6 1 | 5 2 8
------+-------+------
3 9 1 | 6 5 4 | 7 8 2
6 8 7 | 1 2 3 | 9 4 5
5 2 4 | 7 8 9 | 1 3 6
------+-------+------
7 5 9 | 8 3 6 | 2 1 4
1 3 8 | 4 9 2 | 6 5 7
2 4 6 | 5 1 7 | 8 9 3
```

Intermediate Solution 87

```
6 1 9 | 8 5 2 | 7 4 3
3 4 2 | 9 6 7 | 5 8 1
7 8 5 | 4 1 3 | 2 9 6
------+-------+------
8 5 6 | 2 7 4 | 3 1 9
1 3 7 | 5 9 6 | 4 2 8
2 9 4 | 1 3 8 | 6 5 7
------+-------+------
5 7 8 | 3 4 1 | 9 6 2
4 6 1 | 7 2 9 | 8 3 5
9 2 3 | 6 8 5 | 1 7 4
```

Intermediate Solution 88

```
1 5 6 | 8 2 7 | 4 3 9
4 7 3 | 5 6 9 | 1 2 8
2 9 8 | 1 4 3 | 6 5 7
------+-------+------
9 4 7 | 2 8 5 | 3 6 1
3 1 2 | 7 9 6 | 5 8 4
8 6 5 | 3 1 4 | 9 7 2
------+-------+------
5 2 4 | 6 7 1 | 8 9 3
7 3 1 | 9 5 8 | 2 4 6
6 8 9 | 4 3 2 | 7 1 5
```

Intermediate Solution 89

```
8 1 3 | 2 6 9 | 7 4 5
7 2 5 | 8 1 4 | 6 3 9
6 4 9 | 3 5 7 | 8 2 1
------+-------+------
9 3 6 | 5 8 1 | 2 7 4
1 7 4 | 9 2 6 | 3 5 8
2 5 8 | 7 4 3 | 1 9 6
------+-------+------
4 9 2 | 6 7 8 | 5 1 3
5 8 1 | 4 3 2 | 9 6 7
3 6 7 | 1 9 5 | 4 8 2
```

Intermediate Solution 90

```
5 6 1 | 2 9 4 | 8 7 3
2 3 8 | 7 5 1 | 4 6 9
4 7 9 | 8 6 3 | 1 5 2
------+-------+------
3 4 7 | 6 2 9 | 5 1 8
6 1 5 | 4 3 8 | 2 9 7
8 9 2 | 5 1 7 | 6 3 4
------+-------+------
1 2 4 | 9 7 6 | 3 8 5
9 8 3 | 1 4 5 | 7 2 6
7 5 6 | 3 8 2 | 9 4 1
```

Intermediate Solution 91

8	7	2	5	6	9	4	3	1
5	1	4	3	8	2	7	6	9
3	6	9	4	7	1	2	5	8
2	8	5	9	3	4	1	7	6
6	4	1	2	5	7	9	8	3
9	3	7	6	1	8	5	4	2
1	9	3	7	4	6	8	2	5
7	5	8	1	2	3	6	9	4
4	2	6	8	9	5	3	1	7

Intermediate Solution 92

8	7	1	5	9	6	3	4	2
2	4	3	8	7	1	6	9	5
9	5	6	2	4	3	1	7	8
6	3	8	9	5	4	2	1	7
5	2	9	1	6	7	8	3	4
4	1	7	3	2	8	9	5	6
7	8	5	6	1	9	4	2	3
1	6	4	7	3	2	5	8	9
3	9	2	4	8	5	7	6	1

Intermediate Solution 93

8	1	6	3	9	2	5	7	4
5	4	2	6	8	7	3	9	1
3	7	9	4	1	5	6	8	2
7	5	3	1	2	6	8	4	9
2	8	4	5	7	9	1	3	6
9	6	1	8	4	3	2	5	7
6	2	8	9	5	4	7	1	3
1	9	7	2	3	8	4	6	5
4	3	5	7	6	1	9	2	8

Intermediate Solution 94

4	6	2	9	7	5	8	3	1
5	8	3	4	2	1	6	9	7
7	9	1	8	6	3	5	4	2
3	2	8	6	4	9	7	1	5
1	7	4	5	8	2	9	6	3
6	5	9	1	3	7	4	2	8
8	3	5	2	9	4	1	7	6
2	4	6	7	1	8	3	5	9
9	1	7	3	5	6	2	8	4

Intermediate Solution 95

7	8	6	5	4	1	2	9	3
1	2	4	9	3	6	7	5	8
9	3	5	8	7	2	6	4	1
8	9	1	7	6	4	5	3	2
2	6	7	3	1	5	4	8	9
5	4	3	2	9	8	1	7	6
6	7	2	4	8	9	3	1	5
3	5	9	1	2	7	8	6	4
4	1	8	6	5	3	9	2	7

Intermediate Solution 96

4	9	5	1	8	6	2	7	3
3	6	8	7	4	2	9	1	5
7	1	2	3	9	5	6	8	4
8	2	6	9	1	4	5	3	7
9	5	3	8	2	7	4	6	1
1	4	7	5	6	3	8	9	2
2	3	4	6	7	9	1	5	8
6	7	1	2	5	8	3	4	9
5	8	9	4	3	1	7	2	6

Intermediate Solution 97

2	1	8	3	4	7	6	9	5
5	9	7	2	1	6	8	3	4
4	6	3	8	5	9	7	1	2
7	8	4	1	6	3	2	5	9
9	3	1	5	8	2	4	6	7
6	2	5	7	9	4	1	8	3
1	7	9	6	2	5	3	4	8
8	4	2	9	3	1	5	7	6
3	5	6	4	7	8	9	2	1

Intermediate Solution 98

5	9	2	8	1	6	3	7	4
1	6	4	2	3	7	8	9	5
7	8	3	9	5	4	6	1	2
3	4	1	6	7	5	2	8	9
8	5	7	3	9	2	4	6	1
6	2	9	1	4	8	5	3	7
2	7	8	5	6	9	1	4	3
9	1	5	4	8	3	7	2	6
4	3	6	7	2	1	9	5	8

Intermediate Solution 99

4	9	6	3	1	8	5	7	2
8	7	1	9	5	2	4	3	6
3	2	5	4	7	6	8	9	1
1	4	2	6	8	9	3	5	7
9	5	8	2	3	7	1	6	4
6	3	7	1	4	5	2	8	9
2	8	4	7	9	3	6	1	5
7	6	3	5	2	1	9	4	8
5	1	9	8	6	4	7	2	3

Intermediate Solution 100

1	2	3	4	6	7	8	9	5
5	6	9	8	2	1	3	7	4
8	7	4	5	3	9	2	6	1
3	5	6	9	1	4	7	8	2
2	9	8	3	7	5	1	4	6
7	4	1	6	8	2	9	5	3
6	3	7	1	4	8	5	2	9
9	1	2	7	5	6	4	3	8
4	8	5	2	9	3	6	1	7

Intermediate Solution 101

2	1	7	6	5	3	9	8	4
6	5	8	2	4	9	7	1	3
9	3	4	7	1	8	2	5	6
3	2	9	4	7	5	8	6	1
7	8	6	3	9	1	4	2	5
1	4	5	8	6	2	3	7	9
4	7	1	9	8	6	5	3	2
8	6	3	5	2	4	1	9	7
5	9	2	1	3	7	6	4	8

Intermediate Solution 102

1	4	6	5	2	9	3	8	7
9	8	3	4	6	7	1	5	2
2	7	5	3	8	1	9	6	4
5	9	7	2	1	4	8	3	6
3	6	8	9	7	5	2	4	1
4	1	2	8	3	6	7	9	5
6	3	9	7	4	2	5	1	8
8	2	4	1	5	3	6	7	9
7	5	1	6	9	8	4	2	3

Intermediate Solution 103

6	3	1	4	7	5	9	2	8
7	2	5	8	9	6	1	3	4
4	9	8	2	3	1	6	7	5
3	4	9	1	8	7	5	6	2
1	7	6	9	5	2	4	8	3
8	5	2	6	4	3	7	9	1
5	6	7	3	1	8	2	4	9
9	1	3	7	2	4	8	5	6
2	8	4	5	6	9	3	1	7

Intermediate Solution 104

1	8	9	5	3	2	7	4	6
4	6	7	1	8	9	3	5	2
2	5	3	4	6	7	9	1	8
5	2	6	3	1	4	8	9	7
3	9	1	8	7	5	2	6	4
8	7	4	9	2	6	1	3	5
6	1	2	7	4	3	5	8	9
9	4	8	2	5	1	6	7	3
7	3	5	6	9	8	4	2	1

Intermediate Solution 105

9	5	6	4	8	2	1	3	7
1	8	4	3	5	7	6	9	2
2	7	3	6	1	9	8	4	5
8	1	2	7	4	3	5	6	9
3	4	7	5	9	6	2	1	8
6	9	5	8	2	1	3	7	4
5	6	8	1	7	4	9	2	3
7	2	1	9	3	5	4	8	6
4	3	9	2	6	8	7	5	1

Intermediate Solution 106

2	5	7	3	6	8	9	4	1
1	6	8	4	9	5	2	3	7
3	9	4	2	7	1	5	8	6
6	4	1	9	3	2	8	7	5
7	2	3	5	8	4	1	6	9
9	8	5	7	1	6	3	2	4
4	3	9	8	5	7	6	1	2
5	1	2	6	4	3	7	9	8
8	7	6	1	2	9	4	5	3

Intermediate Solution 107

4	7	8	5	1	2	3	6	9
9	6	1	8	3	7	4	2	5
2	3	5	4	6	9	8	7	1
5	2	4	1	8	3	7	9	6
1	9	3	7	2	6	5	4	8
6	8	7	9	4	5	1	3	2
7	1	2	3	9	8	6	5	4
3	4	6	2	5	1	9	8	7
8	5	9	6	7	4	2	1	3

Intermediate Solution 108

5	1	9	6	2	3	4	7	8
8	3	4	5	9	7	6	1	2
2	7	6	4	8	1	3	9	5
6	2	8	1	5	4	7	3	9
1	5	7	2	3	9	8	6	4
9	4	3	8	7	6	5	2	1
4	6	5	3	1	2	9	8	7
7	8	1	9	6	5	2	4	3
3	9	2	7	4	8	1	5	6

Intermediate Solution 109

7	9	1	8	4	6	5	3	2
5	6	4	7	2	3	9	1	8
8	2	3	1	5	9	6	7	4
2	5	7	6	3	8	1	4	9
9	3	6	5	1	4	2	8	7
4	1	8	2	9	7	3	5	6
1	7	2	4	6	5	8	9	3
3	4	5	9	8	2	7	6	1
6	8	9	3	7	1	4	2	5

Intermediate Solution 110

7	4	5	2	1	9	3	6	8
9	3	1	5	6	8	7	2	4
6	2	8	4	7	3	5	9	1
5	9	6	7	3	4	1	8	2
3	1	7	8	2	6	4	5	9
2	8	4	1	9	5	6	3	7
8	6	3	9	4	1	2	7	5
1	5	2	3	8	7	9	4	6
4	7	9	6	5	2	8	1	3

Intermediate Solution 111

4	2	9	5	1	8	6	3	7
7	3	1	2	9	6	5	4	8
8	5	6	3	7	4	9	1	2
2	4	5	1	3	9	8	7	6
3	9	8	4	6	7	2	5	1
6	1	7	8	2	5	4	9	3
5	6	2	7	4	3	1	8	9
1	8	3	9	5	2	7	6	4
9	7	4	6	8	1	3	2	5

Intermediate Solution 112

1	6	4	9	3	5	2	8	7
8	9	3	7	4	2	1	5	6
7	2	5	8	1	6	3	4	9
9	4	2	1	7	8	5	6	3
3	1	6	4	5	9	8	7	2
5	7	8	6	2	3	4	9	1
2	5	7	3	9	4	6	1	8
4	8	9	2	6	1	7	3	5
6	3	1	5	8	7	9	2	4

Intermediate Solution 113

1	4	6	9	5	8	7	2	3
2	5	3	7	4	6	9	1	8
8	9	7	3	1	2	6	4	5
6	3	8	2	9	1	4	5	7
4	7	9	8	3	5	1	6	2
5	2	1	6	7	4	8	3	9
9	1	4	5	8	3	2	7	6
7	6	5	1	2	9	3	8	4
3	8	2	4	6	7	5	9	1

Intermediate Solution 114

5	6	1	2	3	4	9	8	7
8	7	3	1	6	9	5	2	4
2	9	4	7	8	5	3	1	6
7	1	9	6	5	2	4	3	8
3	8	6	4	7	1	2	5	9
4	5	2	8	9	3	6	7	1
9	4	8	3	2	7	1	6	5
1	2	7	5	4	6	8	9	3
6	3	5	9	1	8	7	4	2

Intermediate Solution 115

5	7	6	2	9	1	4	3	8
8	4	2	3	7	6	9	1	5
3	1	9	8	5	4	6	7	2
4	9	1	5	3	2	7	8	6
2	3	8	6	4	7	1	5	9
6	5	7	9	1	8	3	2	4
7	8	4	1	2	9	5	6	3
1	6	5	4	8	3	2	9	7
9	2	3	7	6	5	8	4	1

Intermediate Solution 116

4	3	8	6	1	9	7	5	2
6	2	9	7	4	5	8	3	1
5	7	1	8	3	2	4	9	6
2	8	5	3	7	4	6	1	9
7	1	4	9	2	6	5	8	3
3	9	6	5	8	1	2	7	4
8	5	2	1	6	3	9	4	7
1	6	7	4	9	8	3	2	5
9	4	3	2	5	7	1	6	8

Intermediate Solution 117

7	1	4	9	3	6	2	5	8
9	3	5	2	8	4	6	1	7
8	6	2	5	1	7	9	3	4
5	8	1	3	7	9	4	2	6
4	7	6	1	5	2	3	8	9
3	2	9	4	6	8	5	7	1
1	4	7	6	2	5	8	9	3
2	9	3	8	4	1	7	6	5
6	5	8	7	9	3	1	4	2

Intermediate Solution 118

9	2	8	4	3	5	6	1	7
3	5	6	2	1	7	8	4	9
1	7	4	6	8	9	5	2	3
7	9	2	1	4	8	3	5	6
6	4	5	9	7	3	2	8	1
8	3	1	5	6	2	9	7	4
4	8	9	7	2	6	1	3	5
2	6	7	3	5	1	4	9	8
5	1	3	8	9	4	7	6	2

Intermediate Solution 119

9	3	6	5	1	4	8	7	2
5	1	8	7	2	3	4	9	6
2	4	7	8	9	6	1	3	5
7	5	4	2	6	8	3	1	9
3	6	1	4	5	9	2	8	7
8	9	2	1	3	7	6	5	4
1	7	9	3	4	2	5	6	8
6	2	5	9	8	1	7	4	3
4	8	3	6	7	5	9	2	1

Intermediate Solution 120

6	9	4	3	8	5	1	2	7
1	5	3	2	7	6	9	8	4
2	7	8	1	4	9	6	5	3
4	8	7	9	6	1	2	3	5
3	6	9	8	5	2	4	7	1
5	1	2	7	3	4	8	9	6
7	2	5	6	1	8	3	4	9
9	4	1	5	2	3	7	6	8
8	3	6	4	9	7	5	1	2

Intermediate Solution 121

9	6	3	2	1	8	5	7	4
1	4	8	6	7	5	2	3	9
7	5	2	4	3	9	1	8	6
4	3	1	8	2	6	7	9	5
2	8	9	7	5	4	3	6	1
6	7	5	1	9	3	8	4	2
3	1	4	9	8	2	6	5	7
8	9	7	5	6	1	4	2	3
5	2	6	3	4	7	9	1	8

Intermediate Solution 122

8	3	4	5	6	7	9	1	2
7	5	1	8	9	2	3	4	6
6	2	9	1	4	3	7	8	5
9	1	2	7	3	5	4	6	8
4	7	5	6	8	1	2	9	3
3	8	6	4	2	9	5	7	1
1	4	3	2	7	6	8	5	9
5	9	8	3	1	4	6	2	7
2	6	7	9	5	8	1	3	4

Intermediate Solution 123

4	1	3	6	9	5	2	7	8
5	7	9	8	2	1	4	3	6
6	2	8	3	7	4	9	5	1
9	6	1	7	5	8	3	2	4
3	4	7	2	1	6	8	9	5
8	5	2	9	4	3	1	6	7
1	3	5	4	6	2	7	8	9
2	9	6	1	8	7	5	4	3
7	8	4	5	3	9	6	1	2

Intermediate Solution 124

9	5	4	3	2	8	1	6	7
1	6	2	7	4	5	8	3	9
3	7	8	1	9	6	2	5	4
4	3	7	2	5	1	9	8	6
8	9	6	4	3	7	5	2	1
2	1	5	6	8	9	4	7	3
7	8	9	5	6	4	3	1	2
5	2	1	9	7	3	6	4	8
6	4	3	8	1	2	7	9	5

Intermediate Solution 125

9	5	8	6	1	3	2	7	4
3	7	1	5	2	4	8	9	6
4	6	2	9	8	7	1	5	3
6	8	4	2	3	9	7	1	5
5	3	7	8	6	1	9	4	2
1	2	9	7	4	5	6	3	8
2	4	3	1	7	8	5	6	9
8	1	5	4	9	6	3	2	7
7	9	6	3	5	2	4	8	1

Intermediate Solution 126

9	5	4	7	3	6	8	2	1
6	3	1	8	9	2	7	5	4
7	2	8	5	1	4	6	3	9
8	1	6	2	7	3	9	4	5
3	7	9	6	4	5	2	1	8
5	4	2	9	8	1	3	6	7
4	8	3	1	6	9	5	7	2
2	6	7	4	5	8	1	9	3
1	9	5	3	2	7	4	8	6

Intermediate Solution 127

5	4	1	8	9	6	2	7	3
9	7	8	3	4	2	1	6	5
6	2	3	1	5	7	4	8	9
3	8	9	2	1	4	6	5	7
1	6	4	9	7	5	8	3	2
7	5	2	6	8	3	9	1	4
2	3	5	4	6	8	7	9	1
8	9	7	5	2	1	3	4	6
4	1	6	7	3	9	5	2	8

Intermediate Solution 128

4	8	6	2	7	5	9	3	1
2	9	7	3	8	1	6	4	5
5	3	1	9	4	6	8	2	7
8	4	9	6	1	7	3	5	2
6	7	2	8	5	3	4	1	9
1	5	3	4	2	9	7	8	6
3	6	5	1	9	4	2	7	8
9	1	8	7	3	2	5	6	4
7	2	4	5	6	8	1	9	3

Intermediate Solution 129

6	2	5	4	9	3	8	1	7
8	1	7	6	2	5	4	9	3
3	9	4	1	7	8	6	5	2
2	7	8	5	6	1	3	4	9
1	5	3	9	8	4	2	7	6
9	4	6	2	3	7	1	8	5
4	3	2	7	1	9	5	6	8
7	6	1	8	5	2	9	3	4
5	8	9	3	4	6	7	2	1

Intermediate Solution 130

6	2	9	3	1	4	8	7	5
5	3	7	6	2	8	1	9	4
1	4	8	7	5	9	3	6	2
8	1	3	2	9	6	4	5	7
9	5	2	4	3	7	6	1	8
4	7	6	5	8	1	9	2	3
2	9	1	8	7	3	5	4	6
3	6	5	9	4	2	7	8	1
7	8	4	1	6	5	2	3	9

Intermediate Solution 131

8	5	9	1	7	4	2	3	6
7	2	6	8	3	9	1	4	5
3	1	4	2	5	6	8	7	9
9	3	5	4	6	1	7	2	8
2	4	7	9	8	3	5	6	1
6	8	1	7	2	5	4	9	3
5	7	3	6	1	2	9	8	4
1	9	8	3	4	7	6	5	2
4	6	2	5	9	8	3	1	7

Intermediate Solution 132

4	9	6	1	7	2	3	8	5
7	2	3	4	5	8	1	6	9
8	5	1	6	9	3	4	7	2
6	8	7	5	3	4	9	2	1
9	1	4	8	2	6	5	3	7
2	3	5	9	1	7	6	4	8
5	4	8	2	6	9	7	1	3
1	7	2	3	4	5	8	9	6
3	6	9	7	8	1	2	5	4

Intermediate Solution 133

4	7	3	8	1	6	5	9	2
2	6	5	3	4	9	7	8	1
9	8	1	5	2	7	4	3	6
1	5	8	9	6	2	3	4	7
3	4	2	1	7	5	8	6	9
6	9	7	4	8	3	2	1	5
5	1	9	2	3	8	6	7	4
7	3	4	6	5	1	9	2	8
8	2	6	7	9	4	1	5	3

Intermediate Solution 134

8	7	9	6	1	5	4	2	3
3	1	4	2	8	9	6	5	7
2	6	5	3	4	7	8	9	1
4	3	8	7	2	6	9	1	5
6	2	7	9	5	1	3	8	4
9	5	1	4	3	8	2	7	6
7	8	2	5	6	3	1	4	9
1	9	6	8	7	4	5	3	2
5	4	3	1	9	2	7	6	8

Intermediate Solution 135

4	2	6	8	1	5	3	7	9
5	7	1	2	3	9	8	4	6
8	3	9	4	6	7	5	2	1
6	1	2	3	5	4	9	8	7
9	8	5	6	7	2	1	3	4
3	4	7	1	9	8	6	5	2
2	9	3	7	8	6	4	1	5
1	6	4	5	2	3	7	9	8
7	5	8	9	4	1	2	6	3

Intermediate Solution 136

7	2	9	8	3	5	1	6	4
1	3	6	4	7	9	8	5	2
5	8	4	6	2	1	7	9	3
9	4	5	7	8	3	2	1	6
6	7	3	9	1	2	4	8	5
2	1	8	5	4	6	3	7	9
3	9	7	1	6	4	5	2	8
8	6	2	3	5	7	9	4	1
4	5	1	2	9	8	6	3	7

Intermediate Solution 137

1	4	3	7	5	8	2	9	6
9	7	6	3	2	1	8	5	4
8	2	5	9	6	4	1	7	3
6	5	7	4	9	2	3	1	8
3	9	8	6	1	7	5	4	2
2	1	4	8	3	5	9	6	7
5	6	2	1	7	3	4	8	9
4	3	9	5	8	6	7	2	1
7	8	1	2	4	9	6	3	5

Intermediate Solution 138

4	1	3	9	6	8	2	7	5
2	7	5	1	3	4	9	8	6
6	8	9	2	7	5	4	3	1
9	3	4	5	1	7	8	6	2
1	5	8	4	2	6	3	9	7
7	6	2	3	8	9	5	1	4
3	4	7	8	5	1	6	2	9
5	2	6	7	9	3	1	4	8
8	9	1	6	4	2	7	5	3

Intermediate Solution 139

7	1	8	6	2	5	3	9	4
9	6	2	3	4	1	5	7	8
3	4	5	9	8	7	2	1	6
2	7	6	4	1	8	9	3	5
4	5	9	7	3	2	8	6	1
8	3	1	5	9	6	4	2	7
5	9	3	1	6	4	7	8	2
6	2	7	8	5	9	1	4	3
1	8	4	2	7	3	6	5	9

Intermediate Solution 140

4	3	5	1	8	6	2	7	9
6	8	9	4	2	7	5	3	1
2	7	1	5	9	3	4	8	6
8	2	3	9	1	5	6	4	7
9	4	7	6	3	2	1	5	8
1	5	6	8	7	4	9	2	3
3	6	8	2	5	9	7	1	4
7	9	2	3	4	1	8	6	5
5	1	4	7	6	8	3	9	2

Intermediate Solution 141

3	4	2	6	9	5	1	8	7
6	8	7	3	2	1	4	5	9
1	9	5	4	8	7	2	6	3
8	3	9	1	4	2	5	7	6
7	2	4	5	3	6	8	9	1
5	1	6	8	7	9	3	2	4
4	5	1	7	6	8	9	3	2
9	7	8	2	1	3	6	4	5
2	6	3	9	5	4	7	1	8

Intermediate Solution 142

5	8	4	3	6	7	1	2	9
7	9	2	1	5	8	3	4	6
3	1	6	9	2	4	5	8	7
8	7	5	6	9	2	4	1	3
6	4	1	8	3	5	9	7	2
2	3	9	7	4	1	8	6	5
4	5	8	2	7	9	6	3	1
1	6	7	5	8	3	2	9	4
9	2	3	4	1	6	7	5	8

Intermediate Solution 143

9	2	8	3	1	4	6	7	5
1	6	3	5	8	7	2	4	9
5	4	7	6	9	2	1	8	3
3	7	9	8	2	6	5	1	4
6	5	4	9	7	1	3	2	8
8	1	2	4	3	5	9	6	7
7	9	5	1	6	8	4	3	2
2	3	6	7	4	9	8	5	1
4	8	1	2	5	3	7	9	6

Intermediate Solution 144

6	2	1	4	9	8	7	3	5
3	5	4	6	2	7	1	9	8
8	9	7	3	5	1	4	2	6
9	4	2	1	3	5	8	6	7
5	3	6	8	7	9	2	4	1
1	7	8	2	6	4	9	5	3
2	8	3	7	4	6	5	1	9
4	1	5	9	8	3	6	7	2
7	6	9	5	1	2	3	8	4

Intermediate Solution 145

4	1	5	9	6	8	3	7	2
8	7	3	1	2	5	6	9	4
9	2	6	4	7	3	1	8	5
3	8	7	6	1	4	5	2	9
6	9	2	8	5	7	4	1	3
1	5	4	3	9	2	8	6	7
7	4	8	2	3	6	9	5	1
5	3	1	7	8	9	2	4	6
2	6	9	5	4	1	7	3	8

Intermediate Solution 146

8	5	7	4	2	1	9	3	6
2	3	6	9	7	5	8	4	1
1	4	9	6	8	3	2	5	7
6	8	3	5	4	2	1	7	9
9	7	4	3	1	6	5	8	2
5	1	2	7	9	8	4	6	3
4	2	5	1	6	7	3	9	8
3	6	8	2	5	9	7	1	4
7	9	1	8	3	4	6	2	5

Intermediate Solution 147

9	1	2	4	7	6	5	3	8
3	7	4	5	8	2	1	9	6
5	6	8	9	1	3	4	2	7
8	9	7	1	2	4	3	6	5
6	4	1	3	9	5	7	8	2
2	5	3	8	6	7	9	4	1
4	2	6	7	3	1	8	5	9
1	3	9	2	5	8	6	7	4
7	8	5	6	4	9	2	1	3

Intermediate Solution 148

5	1	6	2	7	4	9	8	3
4	9	2	8	3	5	6	7	1
8	7	3	9	6	1	2	4	5
7	2	5	1	8	9	4	3	6
1	3	4	6	5	7	8	2	9
6	8	9	3	4	2	5	1	7
3	5	8	7	2	6	1	9	4
2	4	1	5	9	3	7	6	8
9	6	7	4	1	8	3	5	2

Intermediate Solution 149

9	3	4	8	6	7	5	1	2
7	5	2	9	4	1	3	6	8
1	6	8	5	3	2	9	7	4
3	1	5	2	7	8	4	9	6
8	4	9	6	1	5	2	3	7
6	2	7	4	9	3	8	5	1
5	9	6	7	8	4	1	2	3
4	7	1	3	2	9	6	8	5
2	8	3	1	5	6	7	4	9

Intermediate Solution 150

7	5	2	4	3	8	6	1	9
8	9	3	5	1	6	2	4	7
6	1	4	9	2	7	5	8	3
3	6	8	7	9	5	4	2	1
2	7	5	1	6	4	3	9	8
9	4	1	2	8	3	7	6	5
5	2	7	8	4	1	9	3	6
1	3	9	6	7	2	8	5	4
4	8	6	3	5	9	1	7	2

Expert Solution 1

3	8	4	9	7	1	2	5	6
7	2	1	5	8	6	3	4	9
5	9	6	3	4	2	1	7	8
2	6	5	4	1	9	8	3	7
4	1	3	7	2	8	9	6	5
8	7	9	6	5	3	4	1	2
6	5	8	2	3	4	7	9	1
9	4	2	1	6	7	5	8	3
1	3	7	8	9	5	6	2	4

Expert Solution 2

3	1	9	5	8	4	6	2	7
4	7	8	3	2	6	1	9	5
6	5	2	7	9	1	4	3	8
2	3	1	8	6	5	9	7	4
9	6	4	1	7	2	5	8	3
5	8	7	9	4	3	2	1	6
1	4	6	2	3	8	7	5	9
8	9	5	4	1	7	3	6	2
7	2	3	6	5	9	8	4	1

Expert Solution 3

7	3	4	8	5	6	1	2	9
1	5	6	3	9	2	7	8	4
9	8	2	4	7	1	6	5	3
6	9	3	2	8	5	4	1	7
4	2	7	9	1	3	8	6	5
8	1	5	6	4	7	9	3	2
2	7	8	5	6	9	3	4	1
5	6	1	7	3	4	2	9	8
3	4	9	1	2	8	5	7	6

Expert Solution 4

5	7	9	4	1	8	6	3	2
3	1	4	6	2	9	5	7	8
8	6	2	3	7	5	4	9	1
7	2	5	9	6	4	8	1	3
9	3	1	5	8	2	7	4	6
6	4	8	1	3	7	9	2	5
4	8	7	2	5	3	1	6	9
1	5	3	7	9	6	2	8	4
2	9	6	8	4	1	3	5	7

Expert Solution 5

4	5	8	1	6	9	7	2	3
2	7	6	3	5	8	1	9	4
3	1	9	7	4	2	8	5	6
6	3	1	9	2	7	5	4	8
5	2	7	4	8	1	6	3	9
9	8	4	6	3	5	2	1	7
8	6	2	5	9	4	3	7	1
7	4	3	2	1	6	9	8	5
1	9	5	8	7	3	4	6	2

Expert Solution 6

9	3	7	8	1	6	5	4	2
2	4	5	9	3	7	6	8	1
6	8	1	2	5	4	7	9	3
8	1	2	7	4	3	9	6	5
3	9	6	5	8	1	2	7	4
5	7	4	6	9	2	1	3	8
4	5	9	1	7	8	3	2	6
1	2	8	3	6	9	4	5	7
7	6	3	4	2	5	8	1	9

Expert Solution 7

1	3	9	8	5	2	7	6	4
4	7	8	6	9	3	2	1	5
2	5	6	1	4	7	9	3	8
5	8	1	7	6	9	3	4	2
3	6	4	5	2	1	8	7	9
9	2	7	3	8	4	6	5	1
7	9	2	4	1	6	5	8	3
6	1	5	2	3	8	4	9	7
8	4	3	9	7	5	1	2	6

Expert Solution 8

6	7	8	3	1	9	5	4	2
2	3	4	6	8	5	1	9	7
1	5	9	4	7	2	6	8	3
9	1	5	2	6	7	4	3	8
8	2	3	1	5	4	7	6	9
4	6	7	8	9	3	2	1	5
7	9	6	5	3	1	8	2	4
3	4	1	7	2	8	9	5	6
5	8	2	9	4	6	3	7	1

Expert Solution 9

2	4	8	6	1	5	9	3	7
3	1	5	9	2	7	4	8	6
6	7	9	8	3	4	5	1	2
9	5	4	3	6	1	2	7	8
7	3	6	4	8	2	1	5	9
8	2	1	5	7	9	3	6	4
4	6	3	1	9	8	7	2	5
5	8	2	7	4	3	6	9	1
1	9	7	2	5	6	8	4	3

Expert Solution 10

5	1	2	3	9	8	7	4	6
9	8	7	2	6	4	5	3	1
6	3	4	5	7	1	8	2	9
3	4	8	1	5	6	2	9	7
2	9	1	8	4	7	3	6	5
7	5	6	9	3	2	1	8	4
8	2	9	4	1	5	6	7	3
1	6	3	7	2	9	4	5	8
4	7	5	6	8	3	9	1	2

Expert Solution 11

4	3	2	5	7	1	6	8	9
6	1	7	9	4	8	3	5	2
5	8	9	2	6	3	1	7	4
2	5	1	7	3	4	8	9	6
3	4	6	8	1	9	5	2	7
9	7	8	6	2	5	4	1	3
1	2	5	4	9	6	7	3	8
8	9	4	3	5	7	2	6	1
7	6	3	1	8	2	9	4	5

Expert Solution 12

3	4	6	9	7	2	5	8	1
1	8	9	3	5	4	7	2	6
7	5	2	6	1	8	9	3	4
4	6	1	5	8	9	3	7	2
5	2	8	7	3	6	4	1	9
9	3	7	4	2	1	6	5	8
2	7	4	1	9	5	8	6	3
8	9	3	2	6	7	1	4	5
6	1	5	8	4	3	2	9	7

Expert Solution 13

4	9	1	8	2	6	7	3	5
6	7	8	4	3	5	1	2	9
2	5	3	7	9	1	6	4	8
1	8	6	3	5	2	4	9	7
9	3	5	6	7	4	8	1	2
7	2	4	1	8	9	5	6	3
5	4	7	9	6	3	2	8	1
8	6	9	2	1	7	3	5	4
3	1	2	5	4	8	9	7	6

Expert Solution 14

1	6	8	9	3	2	5	4	7
5	2	4	7	6	1	3	9	8
3	7	9	8	4	5	1	2	6
4	8	5	6	2	9	7	1	3
6	1	2	3	7	8	9	5	4
7	9	3	5	1	4	6	8	2
9	4	7	1	8	3	2	6	5
2	3	1	4	5	6	8	7	9
8	5	6	2	9	7	4	3	1

Expert Solution 15

3	7	2	4	5	1	8	9	6
5	6	9	8	7	2	3	4	1
1	8	4	3	9	6	5	2	7
4	3	5	1	2	7	6	8	9
6	2	7	9	8	3	1	5	4
8	9	1	6	4	5	7	3	2
7	5	3	2	6	9	4	1	8
9	1	8	7	3	4	2	6	5
2	4	6	5	1	8	9	7	3

Expert Solution 16

5	6	3	9	2	7	8	1	4
4	1	7	6	3	8	5	2	9
2	8	9	5	1	4	3	6	7
7	9	1	2	8	5	6	4	3
3	5	4	1	7	6	2	9	8
8	2	6	4	9	3	1	7	5
1	7	2	3	5	9	4	8	6
6	3	8	7	4	1	9	5	2
9	4	5	8	6	2	7	3	1

Expert Solution 17

2	7	3	4	8	1	6	5	9
4	5	8	6	7	9	3	1	2
9	1	6	3	2	5	7	4	8
1	2	4	8	5	7	9	3	6
6	3	9	2	1	4	5	8	7
7	8	5	9	3	6	1	2	4
8	6	2	1	9	3	4	7	5
5	9	1	7	4	8	2	6	3
3	4	7	5	6	2	8	9	1

Expert Solution 18

8	6	7	5	4	3	1	9	2
5	3	9	2	1	8	7	4	6
2	1	4	9	7	6	5	3	8
4	9	8	1	3	5	2	6	7
6	5	1	4	2	7	3	8	9
3	7	2	6	8	9	4	5	1
9	2	3	8	5	1	6	7	4
7	4	6	3	9	2	8	1	5
1	8	5	7	6	4	9	2	3

Expert Solution 19

4	7	1	5	8	3	9	6	2
9	3	8	2	7	6	1	5	4
2	5	6	9	4	1	7	3	8
7	9	3	6	2	4	8	1	5
8	1	4	3	9	5	2	7	6
6	2	5	7	1	8	4	9	3
3	4	2	1	5	7	6	8	9
5	8	7	4	6	9	3	2	1
1	6	9	8	3	2	5	4	7

Expert Solution 20

4	7	5	9	8	2	1	3	6
6	8	2	3	4	1	7	5	9
3	9	1	5	6	7	4	2	8
8	6	9	4	2	3	5	7	1
5	2	3	7	1	6	9	8	4
1	4	7	8	5	9	2	6	3
9	1	6	2	7	8	3	4	5
2	5	8	1	3	4	6	9	7
7	3	4	6	9	5	8	1	2

Expert Solution 21

1	9	6	8	3	5	4	7	2
2	3	5	7	4	1	6	9	8
7	4	8	6	9	2	3	5	1
8	7	2	4	1	6	9	3	5
4	5	3	2	8	9	1	6	7
6	1	9	5	7	3	8	2	4
9	6	7	1	2	4	5	8	3
5	8	1	3	6	7	2	4	9
3	2	4	9	5	8	7	1	6

Expert Solution 22

7	8	6	1	5	9	4	2	3
1	3	4	8	2	6	9	7	5
9	2	5	4	7	3	8	6	1
3	9	2	6	8	4	1	5	7
4	7	8	5	1	2	6	3	9
6	5	1	9	3	7	2	8	4
2	1	9	7	6	5	3	4	8
8	6	7	3	4	1	5	9	2
5	4	3	2	9	8	7	1	6

Expert Solution 23

3	5	4	2	6	8	7	1	9
7	9	8	3	1	4	6	2	5
1	2	6	9	7	5	4	8	3
6	4	7	1	8	9	3	5	2
9	8	5	4	2	3	1	7	6
2	1	3	7	5	6	8	9	4
8	3	2	5	4	7	9	6	1
4	7	1	6	9	2	5	3	8
5	6	9	8	3	1	2	4	7

Expert Solution 24

4	1	5	6	8	2	3	7	9
3	2	6	9	5	7	8	1	4
7	8	9	4	3	1	6	2	5
1	4	3	7	2	6	9	5	8
5	6	2	8	9	3	7	4	1
9	7	8	1	4	5	2	3	6
8	3	1	2	6	4	5	9	7
6	5	4	3	7	9	1	8	2
2	9	7	5	1	8	4	6	3

Expert Solution 25

8	6	3	9	4	7	5	1	2
7	5	1	6	3	2	9	4	8
4	2	9	8	1	5	3	6	7
9	1	4	2	7	3	6	8	5
5	7	6	1	8	4	2	9	3
2	3	8	5	6	9	1	7	4
3	4	2	7	9	6	8	5	1
1	9	5	4	2	8	7	3	6
6	8	7	3	5	1	4	2	9

Expert Solution 26

7	2	9	3	4	1	5	6	8
1	5	8	2	7	6	9	3	4
4	3	6	8	9	5	7	1	2
3	8	4	7	1	9	6	2	5
9	7	1	5	6	2	8	4	3
5	6	2	4	8	3	1	7	9
8	1	5	6	2	4	3	9	7
2	9	3	1	5	7	4	8	6
6	4	7	9	3	8	2	5	1

Expert Solution 27

6	5	1	8	7	4	9	2	3
9	3	7	5	1	2	4	8	6
2	4	8	6	3	9	5	7	1
4	1	2	3	9	8	6	5	7
5	9	6	7	4	1	2	3	8
7	8	3	2	5	6	1	9	4
1	2	5	4	8	3	7	6	9
8	6	4	9	2	7	3	1	5
3	7	9	1	6	5	8	4	2

Expert Solution 28

7	9	6	2	8	5	3	1	4
5	2	3	7	1	4	8	6	9
1	8	4	9	6	3	5	2	7
6	1	9	8	2	7	4	3	5
2	4	7	5	3	6	9	8	1
8	3	5	1	4	9	2	7	6
3	7	2	4	5	1	6	9	8
4	6	1	3	9	8	7	5	2
9	5	8	6	7	2	1	4	3

Expert Solution 29

4	7	1	5	2	9	8	6	3
3	2	8	6	4	1	7	9	5
9	5	6	3	7	8	1	4	2
7	4	9	2	1	3	6	5	8
6	1	5	7	8	4	3	2	9
2	8	3	9	6	5	4	7	1
8	9	2	4	3	7	5	1	6
1	6	7	8	5	2	9	3	4
5	3	4	1	9	6	2	8	7

Expert Solution 30

4	7	3	6	9	2	5	8	1
5	2	6	8	3	1	4	9	7
9	1	8	5	4	7	6	2	3
2	6	1	4	7	8	9	3	5
3	5	9	2	1	6	8	7	4
8	4	7	9	5	3	2	1	6
6	3	4	7	8	9	1	5	2
1	9	2	3	6	5	7	4	8
7	8	5	1	2	4	3	6	9

Expert Solution 31

5	9	2	4	8	7	1	6	3
3	6	1	9	5	2	4	7	8
7	8	4	3	1	6	9	5	2
9	2	6	8	4	3	5	1	7
4	7	3	5	2	1	6	8	9
8	1	5	6	7	9	2	3	4
6	4	9	1	3	8	7	2	5
1	3	7	2	9	5	8	4	6
2	5	8	7	6	4	3	9	1

Expert Solution 32

5	8	2	4	7	9	3	1	6
4	7	9	6	3	1	2	5	8
3	6	1	8	2	5	9	7	4
8	9	3	1	5	4	7	6	2
1	4	7	2	6	3	5	8	9
6	2	5	7	9	8	1	4	3
7	3	8	5	4	2	6	9	1
2	5	4	9	1	6	8	3	7
9	1	6	3	8	7	4	2	5

Expert Solution 33

3	1	6	7	9	5	8	2	4
7	4	2	1	3	8	9	5	6
5	9	8	2	4	6	1	3	7
2	6	1	5	7	9	3	4	8
4	8	7	3	2	1	6	9	5
9	5	3	8	6	4	2	7	1
1	3	4	9	8	7	5	6	2
8	7	9	6	5	2	4	1	3
6	2	5	4	1	3	7	8	9

Expert Solution 34

1	4	9	3	6	5	7	2	8
3	2	7	1	9	8	4	5	6
8	6	5	4	7	2	1	3	9
7	3	2	6	8	4	9	1	5
5	1	4	9	3	7	6	8	2
6	9	8	2	5	1	3	7	4
9	7	3	5	2	6	8	4	1
2	8	1	7	4	9	5	6	3
4	5	6	8	1	3	2	9	7

Expert Solution 35

5	7	8	2	1	4	9	6	3
1	9	4	6	8	3	7	5	2
2	6	3	5	7	9	4	8	1
9	2	6	8	4	5	3	1	7
3	4	1	7	9	6	5	2	8
7	8	5	3	2	1	6	4	9
4	3	2	1	5	7	8	9	6
8	5	7	9	6	2	1	3	4
6	1	9	4	3	8	2	7	5

Expert Solution 36

6	7	2	1	4	3	9	5	8
3	5	8	9	7	2	4	1	6
1	9	4	6	8	5	2	7	3
8	4	7	5	3	6	1	9	2
9	2	6	4	1	8	7	3	5
5	1	3	2	9	7	8	6	4
2	8	1	3	5	9	6	4	7
7	3	9	8	6	4	5	2	1
4	6	5	7	2	1	3	8	9

Expert Solution 37

2	3	1	7	8	5	4	9	6
4	9	5	1	6	3	7	8	2
7	8	6	4	9	2	5	1	3
9	7	2	6	4	1	8	3	5
8	6	3	5	2	7	1	4	9
5	1	4	8	3	9	6	2	7
6	2	7	3	1	8	9	5	4
3	4	8	9	5	6	2	7	1
1	5	9	2	7	4	3	6	8

Expert Solution 38

8	3	7	6	2	9	1	5	4
5	6	1	7	4	8	9	3	2
9	4	2	1	5	3	6	8	7
4	1	5	2	8	6	3	7	9
3	8	6	9	7	4	5	2	1
2	7	9	3	1	5	8	4	6
7	9	4	5	3	1	2	6	8
6	5	8	4	9	2	7	1	3
1	2	3	8	6	7	4	9	5

Expert Solution 39

6	1	9	3	5	4	8	7	2
3	8	7	1	2	9	5	4	6
4	2	5	7	8	6	9	1	3
2	7	1	6	9	3	4	5	8
8	6	4	5	7	2	3	9	1
5	9	3	8	4	1	6	2	7
1	5	6	9	3	7	2	8	4
9	3	2	4	1	8	7	6	5
7	4	8	2	6	5	1	3	9

Expert Solution 40

3	8	5	4	7	9	2	1	6
6	2	4	1	5	8	3	7	9
1	9	7	6	3	2	4	8	5
8	7	6	2	4	3	5	9	1
9	1	2	8	6	5	7	4	3
5	4	3	9	1	7	8	6	2
4	3	9	7	2	1	6	5	8
7	5	8	3	9	6	1	2	4
2	6	1	5	8	4	9	3	7

Expert Solution 41

6	4	9	8	3	1	5	2	7
7	8	2	9	5	6	1	3	4
3	5	1	2	7	4	6	9	8
4	9	3	5	1	7	8	6	2
8	2	7	6	9	3	4	5	1
1	6	5	4	8	2	3	7	9
5	1	4	7	6	9	2	8	3
9	3	8	1	2	5	7	4	6
2	7	6	3	4	8	9	1	5

Expert Solution 42

1	5	2	9	7	8	3	6	4
9	8	7	6	3	4	2	5	1
4	6	3	1	5	2	9	7	8
8	4	5	3	1	6	7	2	9
6	7	9	4	2	5	8	1	3
3	2	1	8	9	7	5	4	6
2	1	6	7	8	3	4	9	5
5	9	8	2	4	1	6	3	7
7	3	4	5	6	9	1	8	2

Expert Solution 43

9	2	4	3	6	5	7	1	8
8	1	3	9	7	2	6	5	4
7	5	6	1	8	4	3	2	9
4	8	2	6	9	3	1	7	5
3	9	1	4	5	7	8	6	2
6	7	5	2	1	8	9	4	3
1	3	9	5	2	6	4	8	7
5	6	7	8	4	9	2	3	1
2	4	8	7	3	1	5	9	6

Expert Solution 44

9	8	6	2	7	5	4	3	1
5	3	4	8	9	1	2	7	6
2	7	1	6	4	3	8	9	5
8	4	3	9	5	6	7	1	2
6	2	7	3	1	4	9	5	8
1	9	5	7	2	8	3	6	4
7	5	8	4	6	9	1	2	3
4	6	2	1	3	7	5	8	9
3	1	9	5	8	2	6	4	7

Expert Solution 45

7	3	8	6	1	5	2	9	4
5	4	2	9	3	8	6	1	7
9	1	6	2	7	4	8	3	5
2	6	7	3	5	9	4	8	1
1	8	5	4	2	7	9	6	3
4	9	3	1	8	6	7	5	2
3	7	1	8	6	2	5	4	9
8	5	4	7	9	1	3	2	6
6	2	9	5	4	3	1	7	8

Expert Solution 46

6	3	5	4	7	8	9	2	1
4	9	7	1	3	2	5	8	6
1	2	8	6	5	9	3	7	4
2	5	9	8	1	3	6	4	7
7	6	4	9	2	5	1	3	8
8	1	3	7	6	4	2	5	9
5	4	1	3	8	6	7	9	2
9	7	2	5	4	1	8	6	3
3	8	6	2	9	7	4	1	5

Expert Solution 47

7	8	1	9	3	6	5	2	4
5	9	2	8	1	4	7	6	3
3	6	4	7	2	5	1	8	9
6	3	9	5	7	8	2	4	1
2	4	5	1	6	3	9	7	8
1	7	8	2	4	9	6	3	5
8	5	7	4	9	2	3	1	6
9	1	6	3	8	7	4	5	2
4	2	3	6	5	1	8	9	7

Expert Solution 48

3	2	8	6	7	9	5	1	4
7	9	5	3	4	1	8	6	2
1	4	6	2	5	8	3	9	7
8	5	2	4	1	3	6	7	9
4	1	7	9	8	6	2	5	3
9	6	3	5	2	7	4	8	1
6	7	4	8	9	2	1	3	5
5	3	1	7	6	4	9	2	8
2	8	9	1	3	5	7	4	6

Expert Solution 49

7	1	2	8	5	4	6	3	9
9	5	6	2	3	1	4	8	7
8	4	3	9	7	6	2	1	5
1	7	5	3	6	2	8	9	4
4	2	8	1	9	5	7	6	3
3	6	9	4	8	7	1	5	2
2	9	4	5	1	8	3	7	6
5	8	7	6	4	3	9	2	1
6	3	1	7	2	9	5	4	8

Expert Solution 50

8	5	1	7	9	2	4	6	3
7	9	3	4	6	5	2	8	1
4	6	2	8	3	1	5	9	7
2	3	8	6	1	7	9	5	4
1	4	5	9	2	8	3	7	6
6	7	9	3	5	4	1	2	8
5	8	7	2	4	3	6	1	9
3	1	6	5	7	9	8	4	2
9	2	4	1	8	6	7	3	5

Expert Solution 51

9	3	6	2	7	1	5	4	8
2	1	5	6	8	4	3	7	9
7	4	8	5	3	9	1	2	6
1	8	2	7	9	5	4	6	3
5	7	3	4	6	2	8	9	1
6	9	4	8	1	3	2	5	7
4	6	7	3	2	8	9	1	5
8	5	9	1	4	6	7	3	2
3	2	1	9	5	7	6	8	4

Expert Solution 52

8	6	3	9	1	2	7	4	5
5	7	1	8	4	3	2	6	9
2	4	9	7	5	6	1	8	3
7	1	4	6	3	9	8	5	2
3	5	2	1	8	4	6	9	7
9	8	6	2	7	5	4	3	1
4	2	8	5	9	1	3	7	6
6	3	5	4	2	7	9	1	8
1	9	7	3	6	8	5	2	4

Expert Solution 53

1	5	9	6	2	3	8	7	4
7	2	8	4	1	5	3	9	6
6	4	3	9	8	7	1	5	2
8	3	7	5	4	1	2	6	9
2	6	1	8	7	9	5	4	3
4	9	5	2	3	6	7	8	1
9	8	6	1	5	2	4	3	7
3	1	4	7	6	8	9	2	5
5	7	2	3	9	4	6	1	8

Expert Solution 54

6	9	5	8	2	4	3	1	7
8	7	2	5	3	1	6	4	9
3	4	1	6	9	7	2	8	5
9	1	8	2	5	3	4	7	6
7	3	4	1	8	6	5	9	2
5	2	6	7	4	9	1	3	8
4	6	9	3	7	5	8	2	1
2	5	3	9	1	8	7	6	4
1	8	7	4	6	2	9	5	3

Expert Solution 55

3	6	4	1	2	8	7	5	9
7	2	8	9	4	5	1	3	6
9	1	5	7	6	3	4	8	2
6	9	3	4	7	2	8	1	5
8	4	1	3	5	6	9	2	7
5	7	2	8	1	9	3	6	4
2	3	7	6	8	4	5	9	1
4	8	6	5	9	1	2	7	3
1	5	9	2	3	7	6	4	8

Expert Solution 56

5	1	3	6	9	7	4	2	8
2	6	9	8	4	3	1	7	5
4	8	7	1	2	5	6	9	3
6	3	1	7	8	2	5	4	9
8	5	2	9	6	4	7	3	1
9	7	4	3	5	1	8	6	2
7	9	6	2	1	8	3	5	4
1	2	5	4	3	6	9	8	7
3	4	8	5	7	9	2	1	6

Expert Solution 57

6	8	7	2	5	1	4	9	3
2	4	3	6	8	9	7	5	1
5	1	9	4	7	3	6	2	8
1	9	8	3	6	7	2	4	5
3	2	6	5	4	8	9	1	7
7	5	4	9	1	2	3	8	6
9	7	5	8	3	4	1	6	2
4	6	1	7	2	5	8	3	9
8	3	2	1	9	6	5	7	4

Expert Solution 58

5	9	1	7	4	6	8	2	3
3	2	4	8	1	5	6	7	9
8	6	7	2	9	3	1	4	5
6	1	3	9	2	8	4	5	7
4	8	2	1	5	7	9	3	6
9	7	5	3	6	4	2	8	1
7	5	9	4	8	1	3	6	2
1	3	8	6	7	2	5	9	4
2	4	6	5	3	9	7	1	8

Expert Solution 59

3	9	8	6	7	2	4	1	5
4	7	5	9	3	1	2	8	6
2	6	1	4	5	8	9	3	7
8	2	9	7	4	3	5	6	1
6	1	3	2	9	5	7	4	8
5	4	7	8	1	6	3	2	9
9	8	6	5	2	4	1	7	3
1	5	4	3	6	7	8	9	2
7	3	2	1	8	9	6	5	4

Expert Solution 60

8	5	2	7	1	3	4	9	6
6	3	7	4	8	9	1	5	2
1	4	9	6	2	5	3	8	7
2	6	3	1	5	8	9	7	4
7	8	4	3	9	6	5	2	1
5	9	1	2	4	7	8	6	3
4	2	5	8	7	1	6	3	9
3	1	8	9	6	2	7	4	5
9	7	6	5	3	4	2	1	8

Expert Solution 61

5	4	7	1	6	3	2	8	9
3	2	1	9	5	8	6	7	4
8	9	6	7	4	2	5	1	3
9	5	8	3	1	7	4	6	2
1	6	2	8	9	4	7	3	5
7	3	4	5	2	6	1	9	8
6	8	3	4	7	5	9	2	1
2	1	5	6	8	9	3	4	7
4	7	9	2	3	1	8	5	6

Expert Solution 62

8	1	2	5	9	3	6	4	7
6	3	5	7	2	4	9	8	1
9	7	4	1	6	8	5	3	2
3	5	8	4	1	9	2	7	6
2	9	6	3	8	7	1	5	4
7	4	1	2	5	6	8	9	3
1	2	7	9	3	5	4	6	8
5	6	3	8	4	2	7	1	9
4	8	9	6	7	1	3	2	5

Expert Solution 63

1	2	6	5	7	4	8	3	9
7	3	8	9	2	6	5	4	1
9	4	5	1	3	8	7	2	6
3	5	7	2	6	9	1	8	4
8	9	4	3	1	5	2	6	7
6	1	2	8	4	7	3	9	5
4	6	3	7	8	1	9	5	2
2	7	9	6	5	3	4	1	8
5	8	1	4	9	2	6	7	3

Expert Solution 64

5	8	3	4	6	9	1	7	2
4	7	6	1	2	5	3	8	9
9	1	2	3	7	8	5	6	4
7	2	5	6	8	1	4	9	3
8	3	4	9	5	7	6	2	1
1	6	9	2	3	4	8	5	7
3	4	7	8	9	6	2	1	5
2	5	8	7	1	3	9	4	6
6	9	1	5	4	2	7	3	8

Expert Solution 65

5	6	3	8	2	7	4	9	1
7	8	4	9	1	5	3	6	2
1	9	2	6	4	3	7	5	8
8	7	6	2	3	9	5	1	4
4	5	9	1	7	8	6	2	3
3	2	1	4	5	6	8	7	9
2	3	7	5	8	1	9	4	6
9	1	8	7	6	4	2	3	5
6	4	5	3	9	2	1	8	7

Expert Solution 66

9	7	6	2	5	8	1	4	3
1	4	3	7	9	6	2	5	8
5	8	2	1	3	4	6	7	9
7	1	8	3	2	5	9	6	4
6	9	5	4	1	7	3	8	2
3	2	4	8	6	9	5	1	7
2	3	7	6	8	1	4	9	5
8	6	9	5	4	2	7	3	1
4	5	1	9	7	3	8	2	6

Expert Solution 67

7	5	2	9	4	1	6	8	3
6	9	8	3	2	5	7	1	4
4	3	1	8	7	6	2	9	5
5	1	9	4	6	8	3	2	7
3	8	4	7	5	2	9	6	1
2	6	7	1	3	9	4	5	8
9	2	3	5	1	4	8	7	6
8	7	5	6	9	3	1	4	2
1	4	6	2	8	7	5	3	9

Expert Solution 68

9	1	4	2	7	6	3	5	8
5	3	8	1	9	4	2	7	6
2	7	6	8	5	3	4	9	1
8	9	3	4	1	7	5	6	2
1	6	7	3	2	5	8	4	9
4	5	2	6	8	9	1	3	7
7	8	9	5	3	1	6	2	4
6	2	5	7	4	8	9	1	3
3	4	1	9	6	2	7	8	5

Expert Solution 69

5	9	2	7	6	1	3	8	4
1	7	8	2	3	4	9	6	5
6	3	4	9	8	5	2	1	7
7	2	5	1	4	6	8	9	3
4	6	3	5	9	8	7	2	1
9	8	1	3	7	2	4	5	6
3	1	6	8	2	7	5	4	9
2	4	7	6	5	9	1	3	8
8	5	9	4	1	3	6	7	2

Expert Solution 70

2	3	4	8	7	5	9	6	1
8	1	6	4	9	2	5	7	3
5	9	7	6	1	3	2	8	4
3	4	5	2	6	7	1	9	8
6	7	8	9	5	1	3	4	2
1	2	9	3	8	4	7	5	6
7	8	3	1	4	9	6	2	5
4	5	1	7	2	6	8	3	9
9	6	2	5	3	8	4	1	7

Expert Solution 71

3	5	8	4	7	2	6	9	1
6	9	4	1	5	8	7	2	3
2	7	1	6	3	9	8	5	4
7	4	3	2	1	6	5	8	9
9	8	5	3	4	7	1	6	2
1	2	6	8	9	5	4	3	7
8	3	7	9	6	4	2	1	5
5	6	9	7	2	1	3	4	8
4	1	2	5	8	3	9	7	6

Expert Solution 72

3	6	8	4	5	2	9	7	1
9	7	1	8	3	6	2	4	5
5	2	4	1	7	9	3	8	6
1	9	2	6	4	5	8	3	7
6	3	5	9	8	7	4	1	2
8	4	7	2	1	3	6	5	9
7	8	6	3	2	1	5	9	4
4	1	9	5	6	8	7	2	3
2	5	3	7	9	4	1	6	8

Expert Solution 73

8	9	4	6	1	3	2	7	5
6	7	2	9	8	5	1	4	3
5	3	1	2	7	4	6	9	8
9	2	7	5	6	8	4	3	1
1	5	6	4	3	2	7	8	9
4	8	3	1	9	7	5	6	2
2	6	5	3	4	9	8	1	7
3	1	8	7	2	6	9	5	4
7	4	9	8	5	1	3	2	6

Expert Solution 74

5	3	1	9	6	7	4	2	8
4	2	7	1	5	8	3	6	9
9	8	6	4	2	3	5	7	1
2	7	3	8	9	5	1	4	6
6	5	9	7	1	4	2	8	3
1	4	8	6	3	2	7	9	5
8	1	2	3	4	6	9	5	7
7	9	5	2	8	1	6	3	4
3	6	4	5	7	9	8	1	2

Expert Solution 75

9	4	6	7	2	5	3	8	1
5	8	1	6	3	4	2	9	7
3	2	7	1	9	8	5	4	6
1	6	8	3	4	7	9	5	2
2	9	4	8	5	1	7	6	3
7	5	3	9	6	2	8	1	4
8	3	2	5	1	6	4	7	9
6	7	9	4	8	3	1	2	5
4	1	5	2	7	9	6	3	8

Expert Solution 76

5	9	2	7	1	4	6	3	8
3	4	1	8	2	6	5	7	9
6	8	7	5	3	9	1	2	4
1	2	8	4	5	3	9	6	7
4	6	3	9	7	1	8	5	2
9	7	5	6	8	2	3	4	1
2	5	6	1	9	7	4	8	3
7	1	4	3	6	8	2	9	5
8	3	9	2	4	5	7	1	6

Expert Solution 77

3	6	7	8	2	9	1	5	4
8	4	1	3	6	5	9	2	7
2	9	5	1	7	4	3	8	6
9	1	2	4	5	3	6	7	8
6	3	4	7	8	1	2	9	5
7	5	8	2	9	6	4	1	3
5	2	9	6	4	8	7	3	1
4	8	3	9	1	7	5	6	2
1	7	6	5	3	2	8	4	9

Expert Solution 78

2	1	3	7	6	9	5	4	8
4	5	6	1	8	2	3	7	9
9	7	8	5	3	4	1	6	2
6	8	5	3	7	1	2	9	4
3	9	4	2	5	8	7	1	6
7	2	1	4	9	6	8	5	3
5	3	9	8	4	7	6	2	1
8	6	2	9	1	5	4	3	7
1	4	7	6	2	3	9	8	5

Expert Solution 79

5	8	3	1	9	6	7	4	2
1	9	4	7	3	2	8	6	5
2	7	6	4	8	5	9	3	1
7	6	5	2	4	9	1	8	3
8	4	1	5	6	3	2	9	7
3	2	9	8	7	1	6	5	4
6	3	7	9	2	4	5	1	8
9	5	2	3	1	8	4	7	6
4	1	8	6	5	7	3	2	9

Expert Solution 80

6	8	9	3	2	4	5	7	1
5	2	4	1	7	8	3	9	6
3	7	1	9	5	6	4	8	2
2	4	6	7	1	3	9	5	8
1	3	5	8	4	9	6	2	7
8	9	7	2	6	5	1	3	4
7	5	8	6	9	1	2	4	3
9	6	3	4	8	2	7	1	5
4	1	2	5	3	7	8	6	9

Expert Solution 81

5	2	7	4	9	8	3	1	6
9	1	4	3	6	5	2	8	7
8	3	6	7	2	1	9	5	4
2	6	3	8	7	9	1	4	5
4	9	5	6	1	3	7	2	8
1	7	8	5	4	2	6	9	3
6	5	1	9	8	7	4	3	2
3	4	2	1	5	6	8	7	9
7	8	9	2	3	4	5	6	1

Expert Solution 82

4	8	1	3	2	7	6	9	5
5	2	6	1	9	8	4	7	3
7	9	3	4	6	5	8	2	1
8	3	5	7	4	9	1	6	2
9	1	2	6	8	3	5	4	7
6	7	4	5	1	2	3	8	9
3	5	8	9	7	4	2	1	6
1	4	9	2	3	6	7	5	8
2	6	7	8	5	1	9	3	4

Expert Solution 83

9	1	5	6	2	3	4	7	8
4	8	2	9	1	7	3	5	6
6	7	3	8	5	4	9	2	1
5	3	1	7	6	8	2	4	9
8	4	7	2	9	1	6	3	5
2	9	6	4	3	5	8	1	7
1	5	4	3	8	6	7	9	2
3	6	9	5	7	2	1	8	4
7	2	8	1	4	9	5	6	3

Expert Solution 84

5	8	3	4	6	2	1	9	7
1	7	4	9	5	8	6	2	3
2	6	9	7	3	1	8	5	4
7	9	5	1	4	3	2	8	6
6	2	8	5	9	7	4	3	1
3	4	1	2	8	6	9	7	5
8	5	2	6	7	4	3	1	9
9	1	6	3	2	5	7	4	8
4	3	7	8	1	9	5	6	2

Expert Solution 85

8	2	1	6	9	7	5	3	4
3	4	6	5	1	2	9	8	7
9	5	7	4	3	8	2	6	1
5	1	8	3	7	9	4	2	6
6	3	4	2	8	5	1	7	9
2	7	9	1	4	6	8	5	3
1	9	5	8	6	3	7	4	2
4	8	3	7	2	1	6	9	5
7	6	2	9	5	4	3	1	8

Expert Solution 86

9	2	6	3	8	7	5	1	4
5	4	8	2	1	6	7	3	9
3	1	7	9	5	4	6	8	2
8	5	4	6	3	9	1	2	7
2	6	3	1	7	5	4	9	8
7	9	1	4	2	8	3	5	6
1	7	2	8	4	3	9	6	5
4	8	9	5	6	1	2	7	3
6	3	5	7	9	2	8	4	1

Expert Solution 87

8	7	5	3	4	6	2	9	1
6	9	4	1	7	2	5	3	8
2	1	3	5	8	9	7	6	4
3	4	6	9	2	8	1	7	5
9	5	7	6	1	4	3	8	2
1	8	2	7	3	5	6	4	9
4	6	1	2	9	7	8	5	3
7	3	9	8	5	1	4	2	6
5	2	8	4	6	3	9	1	7

Expert Solution 88

4	8	3	9	6	5	7	1	2
9	5	1	2	7	4	6	8	3
2	6	7	8	3	1	5	4	9
3	1	8	6	4	7	9	2	5
5	4	9	3	8	2	1	6	7
6	7	2	5	1	9	8	3	4
1	2	4	7	5	8	3	9	6
7	9	6	1	2	3	4	5	8
8	3	5	4	9	6	2	7	1

Expert Solution 89

7	6	2	9	3	1	5	8	4
3	5	4	2	8	6	7	1	9
9	1	8	4	5	7	3	6	2
2	8	7	5	9	3	1	4	6
5	9	6	1	7	4	8	2	3
1	4	3	8	6	2	9	7	5
4	3	1	7	2	5	6	9	8
6	7	9	3	4	8	2	5	1
8	2	5	6	1	9	4	3	7

Expert Solution 90

4	8	3	7	5	9	2	6	1
6	9	1	8	2	3	5	7	4
7	2	5	1	4	6	9	3	8
1	3	2	6	7	5	8	4	9
9	7	6	4	1	8	3	5	2
8	5	4	9	3	2	7	1	6
2	4	7	5	9	1	6	8	3
5	6	9	3	8	4	1	2	7
3	1	8	2	6	7	4	9	5

Expert Solution 91

4	8	9	6	3	2	1	7	5
1	5	3	8	7	4	9	2	6
6	2	7	9	1	5	4	8	3
5	4	1	2	8	9	6	3	7
9	3	2	7	6	1	8	5	4
7	6	8	5	4	3	2	9	1
3	7	6	4	2	8	5	1	9
2	1	5	3	9	6	7	4	8
8	9	4	1	5	7	3	6	2

Expert Solution 92

2	8	9	7	3	4	5	6	1
4	6	7	2	1	5	8	9	3
1	3	5	9	6	8	7	2	4
7	1	8	4	5	2	9	3	6
5	4	6	3	8	9	2	1	7
9	2	3	1	7	6	4	8	5
8	5	4	6	9	3	1	7	2
6	9	1	5	2	7	3	4	8
3	7	2	8	4	1	6	5	9

Expert Solution 93

5	7	9	4	6	1	2	8	3
1	8	3	2	9	5	7	4	6
2	4	6	8	7	3	1	5	9
8	5	7	3	2	9	6	1	4
9	2	1	7	4	6	5	3	8
3	6	4	5	1	8	9	7	2
7	3	5	6	8	2	4	9	1
4	9	2	1	3	7	8	6	5
6	1	8	9	5	4	3	2	7

Expert Solution 94

4	5	6	8	7	2	3	1	9
7	3	9	1	6	5	4	8	2
8	1	2	9	3	4	6	7	5
9	8	3	7	2	1	5	4	6
1	7	5	4	9	6	2	3	8
2	6	4	5	8	3	1	9	7
6	9	1	3	5	8	7	2	4
3	2	7	6	4	9	8	5	1
5	4	8	2	1	7	9	6	3

Expert Solution 95

5	4	8	6	7	1	9	3	2
1	2	6	9	8	3	5	7	4
3	7	9	4	5	2	8	6	1
9	1	7	5	4	8	3	2	6
6	3	2	7	1	9	4	8	5
8	5	4	2	3	6	1	9	7
4	9	3	1	6	7	2	5	8
2	6	1	8	9	5	7	4	3
7	8	5	3	2	4	6	1	9

Expert Solution 96

7	8	2	6	3	1	9	5	4
9	4	6	2	7	5	8	3	1
5	3	1	9	8	4	2	6	7
3	6	8	1	5	2	4	7	9
1	2	9	7	4	3	5	8	6
4	7	5	8	9	6	3	1	2
8	1	7	5	2	9	6	4	3
6	9	3	4	1	8	7	2	5
2	5	4	3	6	7	1	9	8

Expert Solution 97

7	2	8	5	6	9	1	4	3
1	5	3	2	8	4	7	9	6
4	9	6	3	1	7	5	2	8
9	6	2	4	7	8	3	1	5
8	1	5	6	2	3	9	7	4
3	7	4	9	5	1	8	6	2
2	8	9	1	3	6	4	5	7
6	3	1	7	4	5	2	8	9
5	4	7	8	9	2	6	3	1

Expert Solution 98

8	1	3	2	7	5	4	9	6
6	9	5	1	3	4	2	7	8
4	7	2	6	9	8	1	5	3
7	8	1	4	5	3	6	2	9
2	5	9	8	1	6	3	4	7
3	6	4	7	2	9	5	8	1
9	4	8	3	6	2	7	1	5
5	3	7	9	4	1	8	6	2
1	2	6	5	8	7	9	3	4

Expert Solution 99

9	3	6	4	8	1	2	7	5
1	4	2	7	5	6	9	8	3
8	5	7	9	3	2	6	4	1
5	7	8	2	1	3	4	9	6
3	9	1	6	4	8	7	5	2
2	6	4	5	7	9	1	3	8
7	8	3	1	2	4	5	6	9
6	2	5	3	9	7	8	1	4
4	1	9	8	6	5	3	2	7

Expert Solution 100

1	9	2	8	6	5	4	3	7
8	7	5	4	3	1	2	6	9
4	3	6	9	2	7	5	8	1
6	1	3	7	5	4	8	9	2
2	8	7	6	1	9	3	4	5
9	5	4	3	8	2	7	1	6
7	6	1	2	4	3	9	5	8
5	4	9	1	7	8	6	2	3
3	2	8	5	9	6	1	7	4

Expert Solution 101

5	7	9	6	8	4	2	1	3
8	4	2	3	9	1	6	7	5
3	6	1	5	7	2	8	4	9
7	1	4	2	6	9	3	5	8
6	3	8	4	5	7	1	9	2
2	9	5	8	1	3	7	6	4
1	8	3	9	4	6	5	2	7
9	2	6	7	3	5	4	8	1
4	5	7	1	2	8	9	3	6

Expert Solution 102

9	8	5	2	1	4	3	7	6
3	4	1	6	7	8	2	9	5
7	6	2	9	3	5	4	8	1
1	3	8	7	4	6	9	5	2
5	9	4	3	2	1	7	6	8
6	2	7	5	8	9	1	3	4
2	1	3	8	6	7	5	4	9
8	7	9	4	5	2	6	1	3
4	5	6	1	9	3	8	2	7

Expert Solution 103

7	9	6	3	5	2	8	1	4
5	8	3	4	7	1	9	2	6
1	4	2	6	9	8	7	3	5
8	2	9	5	6	7	1	4	3
3	5	4	8	1	9	6	7	2
6	7	1	2	4	3	5	8	9
2	3	7	9	8	5	4	6	1
9	6	8	1	3	4	2	5	7
4	1	5	7	2	6	3	9	8

Expert Solution 104

7	5	8	9	4	6	1	2	3
6	2	9	1	3	8	5	7	4
4	1	3	2	7	5	6	8	9
1	7	5	8	2	9	4	3	6
9	8	4	7	6	3	2	5	1
3	6	2	4	5	1	7	9	8
2	9	6	5	8	4	3	1	7
8	3	7	6	1	2	9	4	5
5	4	1	3	9	7	8	6	2

Expert Solution 105

8	4	3	5	2	7	1	6	9
7	6	5	1	9	4	2	3	8
2	9	1	3	8	6	4	5	7
5	2	6	8	7	3	9	1	4
4	8	7	9	1	5	3	2	6
3	1	9	4	6	2	8	7	5
6	3	2	7	4	8	5	9	1
1	7	4	2	5	9	6	8	3
9	5	8	6	3	1	7	4	2

Expert Solution 106

6	3	9	4	5	2	1	8	7
1	4	8	7	6	3	9	2	5
7	5	2	1	9	8	3	4	6
8	9	4	5	7	1	2	6	3
5	1	7	3	2	6	4	9	8
2	6	3	9	8	4	7	5	1
3	2	1	6	4	5	8	7	9
4	7	5	8	1	9	6	3	2
9	8	6	2	3	7	5	1	4

Expert Solution 107

4	5	9	1	3	6	2	7	8
1	7	8	9	5	2	6	4	3
2	6	3	8	7	4	1	5	9
3	9	6	5	1	8	4	2	7
5	4	7	2	6	3	9	8	1
8	2	1	7	4	9	5	3	6
7	3	5	6	2	1	8	9	4
9	1	4	3	8	5	7	6	2
6	8	2	4	9	7	3	1	5

Expert Solution 108

8	6	7	2	1	4	3	9	5
9	4	2	6	5	3	1	7	8
1	3	5	7	8	9	2	4	6
7	2	4	8	3	6	5	1	9
6	8	1	5	9	2	4	3	7
5	9	3	1	4	7	8	6	2
3	7	6	4	2	5	9	8	1
4	5	8	9	7	1	6	2	3
2	1	9	3	6	8	7	5	4

Expert Solution 109

1	3	6	9	8	7	5	2	4
7	5	8	2	3	4	9	1	6
4	2	9	1	5	6	8	3	7
5	7	1	3	4	2	6	9	8
2	6	4	8	7	9	1	5	3
9	8	3	5	6	1	4	7	2
3	1	5	6	2	8	7	4	9
6	9	7	4	1	3	2	8	5
8	4	2	7	9	5	3	6	1

Expert Solution 110

8	1	6	5	2	4	9	7	3
4	3	2	7	9	8	5	1	6
9	7	5	6	3	1	2	4	8
6	2	4	9	1	3	7	8	5
7	8	1	4	6	5	3	9	2
5	9	3	2	8	7	4	6	1
2	6	7	8	5	9	1	3	4
3	4	8	1	7	2	6	5	9
1	5	9	3	4	6	8	2	7

Expert Solution 111

1	2	7	4	5	8	6	9	3
9	6	5	1	2	3	4	7	8
8	4	3	9	7	6	1	5	2
6	5	4	7	8	9	3	2	1
2	9	1	5	3	4	7	8	6
3	7	8	2	6	1	9	4	5
4	8	9	3	1	2	5	6	7
5	1	2	6	9	7	8	3	4
7	3	6	8	4	5	2	1	9

Expert Solution 112

5	7	2	9	6	1	3	4	8
9	1	6	4	3	8	2	7	5
8	4	3	2	7	5	6	9	1
2	3	4	8	5	7	9	1	6
6	8	1	3	2	9	4	5	7
7	9	5	6	1	4	8	3	2
1	6	9	5	8	3	7	2	4
4	2	7	1	9	6	5	8	3
3	5	8	7	4	2	1	6	9

Expert Solution 113

2	3	9	6	1	4	7	5	8
6	7	8	9	5	2	1	4	3
5	4	1	3	7	8	2	6	9
9	1	2	7	8	5	6	3	4
8	6	7	4	3	9	5	1	2
4	5	3	1	2	6	9	8	7
7	2	6	5	4	3	8	9	1
3	8	5	2	9	1	4	7	6
1	9	4	8	6	7	3	2	5

Expert Solution 114

9	8	4	2	6	3	1	7	5
7	1	5	9	4	8	3	6	2
6	2	3	5	7	1	8	9	4
5	4	6	7	8	9	2	1	3
1	9	2	3	5	6	7	4	8
8	3	7	1	2	4	9	5	6
3	5	8	6	1	7	4	2	9
2	7	9	4	3	5	6	8	1
4	6	1	8	9	2	5	3	7

Expert Solution 115

5	6	1	8	7	3	9	4	2
7	3	2	4	6	9	8	5	1
9	8	4	1	2	5	7	6	3
2	4	9	6	1	8	3	7	5
8	7	5	3	9	4	1	2	6
3	1	6	2	5	7	4	9	8
6	2	7	9	8	1	5	3	4
4	5	8	7	3	2	6	1	9
1	9	3	5	4	6	2	8	7

Expert Solution 116

5	8	7	1	9	6	3	2	4
6	4	1	3	7	2	8	5	9
2	9	3	5	4	8	6	1	7
4	1	9	8	3	7	5	6	2
3	5	2	9	6	4	7	8	1
8	7	6	2	1	5	9	4	3
7	3	8	6	2	1	4	9	5
9	2	5	4	8	3	1	7	6
1	6	4	7	5	9	2	3	8

Expert Solution 117

5	9	1	4	2	7	6	3	8
3	6	2	1	8	5	4	7	9
4	7	8	6	3	9	5	2	1
7	8	6	9	4	2	3	1	5
9	1	5	3	6	8	2	4	7
2	3	4	7	5	1	8	9	6
6	4	7	5	1	3	9	8	2
8	5	9	2	7	4	1	6	3
1	2	3	8	9	6	7	5	4

Expert Solution 118

3	4	9	6	7	2	1	5	8
8	7	1	3	5	9	6	4	2
2	5	6	4	1	8	7	9	3
6	8	2	7	3	4	5	1	9
5	9	4	2	8	1	3	6	7
7	1	3	5	9	6	8	2	4
4	3	5	1	2	7	9	8	6
1	6	8	9	4	3	2	7	5
9	2	7	8	6	5	4	3	1

Expert Solution 119

5	7	3	9	2	4	1	8	6
8	6	9	5	1	3	7	4	2
2	1	4	8	7	6	9	3	5
1	2	7	3	6	5	8	9	4
9	5	8	7	4	1	6	2	3
3	4	6	2	8	9	5	7	1
4	8	5	1	3	7	2	6	9
6	9	2	4	5	8	3	1	7
7	3	1	6	9	2	4	5	8

Expert Solution 120

8	2	5	7	3	6	4	1	9
4	1	3	8	2	9	5	7	6
9	6	7	1	4	5	2	3	8
2	5	9	4	6	7	1	8	3
3	7	1	5	9	8	6	2	4
6	4	8	2	1	3	7	9	5
1	9	4	6	8	2	3	5	7
5	8	2	3	7	4	9	6	1
7	3	6	9	5	1	8	4	2

Expert Solution 121

5	6	2	1	8	7	9	3	4
3	4	1	9	2	6	5	7	8
7	8	9	3	5	4	1	6	2
6	7	4	8	9	3	2	1	5
1	2	8	6	4	5	3	9	7
9	5	3	2	7	1	4	8	6
4	9	6	7	3	2	8	5	1
2	3	7	5	1	8	6	4	9
8	1	5	4	6	9	7	2	3

Expert Solution 122

4	3	8	6	7	5	2	9	1
5	7	9	2	8	1	6	3	4
2	1	6	4	3	9	8	7	5
1	6	4	8	5	3	9	2	7
3	8	7	1	9	2	5	4	6
9	2	5	7	4	6	3	1	8
6	9	1	5	2	4	7	8	3
8	4	2	3	6	7	1	5	9
7	5	3	9	1	8	4	6	2

Expert Solution 123

3	6	8	5	7	4	1	9	2
5	1	9	8	2	6	4	7	3
7	2	4	3	1	9	8	6	5
2	8	3	9	6	1	7	5	4
4	7	5	2	3	8	9	1	6
1	9	6	7	4	5	3	2	8
9	4	1	6	5	3	2	8	7
8	5	2	4	9	7	6	3	1
6	3	7	1	8	2	5	4	9

Expert Solution 124

8	2	5	9	3	4	1	7	6
7	6	9	1	2	8	4	5	3
4	1	3	6	5	7	2	9	8
9	7	1	3	8	6	5	2	4
6	4	8	5	1	2	7	3	9
3	5	2	7	4	9	8	6	1
2	3	4	8	9	5	6	1	7
5	9	7	4	6	1	3	8	2
1	8	6	2	7	3	9	4	5

Expert Solution 125

2	9	7	6	3	8	4	1	5
4	1	3	7	2	5	6	9	8
5	8	6	1	4	9	2	3	7
6	4	5	8	1	7	9	2	3
7	3	1	5	9	2	8	4	6
9	2	8	3	6	4	5	7	1
8	6	2	9	7	1	3	5	4
3	7	9	4	5	6	1	8	2
1	5	4	2	8	3	7	6	9

Expert Solution 126

7	5	2	6	4	8	9	1	3
4	9	1	3	2	5	7	6	8
3	8	6	7	9	1	2	5	4
9	3	5	1	8	2	4	7	6
2	6	7	5	3	4	1	8	9
8	1	4	9	6	7	3	2	5
6	2	3	8	7	9	5	4	1
5	4	8	2	1	3	6	9	7
1	7	9	4	5	6	8	3	2

Expert Solution 127

5	1	2	3	8	9	6	7	4
9	7	4	5	1	6	2	8	3
6	3	8	7	4	2	1	9	5
1	8	3	4	6	7	5	2	9
2	5	6	8	9	1	4	3	7
4	9	7	2	5	3	8	1	6
8	2	5	9	7	4	3	6	1
3	6	9	1	2	5	7	4	8
7	4	1	6	3	8	9	5	2

Expert Solution 128

8	9	4	5	2	3	1	6	7
6	3	5	1	9	7	8	4	2
2	1	7	8	4	6	9	5	3
4	5	3	6	7	1	2	9	8
9	2	6	4	3	8	7	1	5
1	7	8	2	5	9	4	3	6
3	4	2	7	1	5	6	8	9
7	6	9	3	8	4	5	2	1
5	8	1	9	6	2	3	7	4

Expert Solution 129

3	6	1	4	8	5	7	9	2
7	8	5	9	2	3	4	1	6
9	4	2	6	7	1	3	8	5
6	7	9	8	3	4	2	5	1
5	2	3	7	1	6	9	4	8
4	1	8	2	5	9	6	3	7
1	9	6	5	4	7	8	2	3
8	3	7	1	9	2	5	6	4
2	5	4	3	6	8	1	7	9

Expert Solution 130

5	8	4	6	3	2	9	1	7
7	9	2	1	8	4	5	6	3
3	1	6	5	9	7	4	2	8
8	2	9	3	1	6	7	4	5
4	6	7	8	2	5	1	3	9
1	3	5	4	7	9	6	8	2
9	4	3	2	5	1	8	7	6
6	7	8	9	4	3	2	5	1
2	5	1	7	6	8	3	9	4

Expert Solution 131

2	4	9	5	7	6	1	3	8
1	6	7	3	4	8	5	2	9
5	3	8	2	9	1	6	4	7
4	5	6	1	2	7	9	8	3
9	8	1	4	5	3	2	7	6
7	2	3	8	6	9	4	5	1
6	7	5	9	8	2	3	1	4
8	1	4	6	3	5	7	9	2
3	9	2	7	1	4	8	6	5

Expert Solution 132

4	9	7	8	2	5	1	3	6
5	2	6	7	3	1	8	4	9
1	8	3	6	9	4	2	5	7
6	1	8	3	4	7	9	2	5
2	5	4	9	8	6	7	1	3
7	3	9	1	5	2	4	6	8
8	4	5	2	7	3	6	9	1
9	6	2	5	1	8	3	7	4
3	7	1	4	6	9	5	8	2

Expert Solution 133

3	1	8	2	9	6	5	4	7
2	9	4	5	3	7	8	1	6
7	5	6	1	8	4	9	2	3
5	8	7	3	6	2	1	9	4
9	6	3	8	4	1	2	7	5
4	2	1	9	7	5	3	6	8
1	7	2	4	5	3	6	8	9
6	3	9	7	1	8	4	5	2
8	4	5	6	2	9	7	3	1

Expert Solution 134

9	7	2	6	3	1	5	8	4
1	6	8	9	4	5	7	3	2
4	5	3	8	2	7	9	1	6
8	3	7	5	9	4	2	6	1
2	4	9	1	6	8	3	7	5
6	1	5	3	7	2	8	4	9
7	8	6	4	5	9	1	2	3
5	2	4	7	1	3	6	9	8
3	9	1	2	8	6	4	5	7

Expert Solution 135

5	1	2	9	3	4	8	7	6
7	6	3	2	5	8	9	1	4
4	8	9	7	1	6	5	3	2
1	9	4	8	2	7	3	6	5
3	2	5	1	6	9	7	4	8
8	7	6	3	4	5	2	9	1
6	4	8	5	7	3	1	2	9
9	3	1	6	8	2	4	5	7
2	5	7	4	9	1	6	8	3

Expert Solution 136

1	4	7	2	9	6	3	8	5
9	6	3	1	5	8	4	7	2
5	2	8	7	3	4	9	6	1
2	3	5	6	4	9	7	1	8
8	9	4	5	7	1	6	2	3
6	7	1	3	8	2	5	9	4
7	1	2	4	6	5	8	3	9
4	8	6	9	2	3	1	5	7
3	5	9	8	1	7	2	4	6

Expert Solution 137

6	4	3	8	5	9	7	1	2
7	1	2	6	4	3	8	5	9
9	8	5	7	1	2	4	6	3
5	3	8	9	2	4	1	7	6
2	9	7	1	6	8	5	3	4
1	6	4	3	7	5	2	9	8
4	7	6	2	9	1	3	8	5
3	2	9	5	8	7	6	4	1
8	5	1	4	3	6	9	2	7

Expert Solution 138

9	2	6	7	8	1	5	3	4
5	1	7	3	9	4	2	8	6
3	4	8	6	2	5	1	7	9
1	7	3	5	4	9	6	2	8
6	5	9	8	7	2	3	4	1
2	8	4	1	3	6	7	9	5
4	3	1	2	5	8	9	6	7
8	6	2	9	1	7	4	5	3
7	9	5	4	6	3	8	1	2

Expert Solution 139

2	9	3	6	5	1	8	7	4
6	8	7	2	9	4	1	3	5
4	5	1	7	8	3	6	9	2
3	6	8	4	1	5	9	2	7
7	2	9	3	6	8	4	5	1
1	4	5	9	2	7	3	8	6
5	3	2	1	4	9	7	6	8
9	1	6	8	7	2	5	4	3
8	7	4	5	3	6	2	1	9

Expert Solution 140

1	8	2	9	4	5	6	7	3
9	4	6	2	3	7	8	1	5
3	7	5	8	1	6	9	4	2
8	2	7	3	6	4	1	5	9
5	1	3	7	9	8	4	2	6
4	6	9	5	2	1	7	3	8
7	9	4	6	5	3	2	8	1
2	3	1	4	8	9	5	6	7
6	5	8	1	7	2	3	9	4

Expert Solution 141

6	1	2	9	5	7	3	4	8
4	3	9	2	8	6	7	1	5
7	8	5	3	4	1	6	9	2
8	4	3	5	2	9	1	7	6
5	7	1	4	6	8	9	2	3
2	9	6	1	7	3	5	8	4
9	2	4	7	3	5	8	6	1
1	5	8	6	9	4	2	3	7
3	6	7	8	1	2	4	5	9

Expert Solution 142

8	4	2	6	9	3	7	5	1
6	1	9	8	7	5	4	2	3
7	5	3	1	4	2	9	8	6
9	2	7	4	6	1	5	3	8
5	3	4	9	2	8	6	1	7
1	6	8	5	3	7	2	4	9
4	8	1	7	5	9	3	6	2
2	7	6	3	8	4	1	9	5
3	9	5	2	1	6	8	7	4

Expert Solution 143

2	6	5	4	1	9	7	8	3
4	8	9	3	6	7	2	5	1
1	7	3	8	5	2	4	9	6
5	3	8	1	2	6	9	4	7
7	1	6	9	8	4	3	2	5
9	4	2	5	7	3	1	6	8
3	2	1	6	4	8	5	7	9
8	9	7	2	3	5	6	1	4
6	5	4	7	9	1	8	3	2

Expert Solution 144

1	2	8	5	4	6	3	7	9
9	7	3	2	8	1	6	4	5
6	4	5	9	7	3	2	8	1
5	3	4	6	2	9	7	1	8
2	1	7	4	5	8	9	6	3
8	9	6	3	1	7	4	5	2
4	6	1	8	9	2	5	3	7
7	5	2	1	3	4	8	9	6
3	8	9	7	6	5	1	2	4

Expert Solution 145

7	6	1	5	2	3	4	8	9
8	4	3	9	6	7	1	5	2
5	2	9	4	1	8	6	7	3
9	3	7	6	8	4	2	1	5
2	1	8	7	5	9	3	6	4
6	5	4	1	3	2	8	9	7
1	7	2	3	9	6	5	4	8
3	9	5	8	4	1	7	2	6
4	8	6	2	7	5	9	3	1

Expert Solution 146

1	4	2	8	9	6	3	5	7
3	8	6	7	1	5	2	4	9
5	9	7	3	4	2	6	1	8
4	1	9	6	2	8	7	3	5
6	5	8	1	7	3	9	2	4
7	2	3	9	5	4	8	6	1
8	6	5	4	3	9	1	7	2
9	7	4	2	6	1	5	8	3
2	3	1	5	8	7	4	9	6

Expert Solution 147

8	3	4	5	6	1	7	9	2
5	9	7	3	2	4	1	6	8
2	1	6	9	7	8	3	5	4
6	7	5	2	8	9	4	1	3
4	2	1	7	5	3	9	8	6
9	8	3	4	1	6	2	7	5
7	6	8	1	3	2	5	4	9
1	4	2	8	9	5	6	3	7
3	5	9	6	4	7	8	2	1

Expert Solution 148

4	8	6	2	9	5	7	1	3
2	7	3	8	4	1	6	5	9
1	9	5	3	6	7	8	4	2
6	1	8	7	3	9	5	2	4
9	5	7	1	2	4	3	8	6
3	4	2	5	8	6	1	9	7
8	2	9	6	5	3	4	7	1
5	6	1	4	7	2	9	3	8
7	3	4	9	1	8	2	6	5

Expert Solution 149

7	1	2	6	8	3	9	5	4
3	8	5	7	9	4	2	1	6
4	6	9	5	1	2	3	7	8
1	3	8	2	6	9	7	4	5
2	9	4	8	7	5	6	3	1
5	7	6	3	4	1	8	9	2
9	5	7	1	2	8	4	6	3
8	4	3	9	5	6	1	2	7
6	2	1	4	3	7	5	8	9

Expert Solution 150

1	8	5	6	4	7	3	2	9
2	4	6	3	8	9	5	7	1
9	3	7	2	5	1	8	4	6
5	7	3	4	1	2	6	9	8
8	9	1	7	6	3	4	5	2
6	2	4	8	9	5	1	3	7
7	1	2	5	3	6	9	8	4
4	5	9	1	7	8	2	6	3
3	6	8	9	2	4	7	1	5

Expert Solution 151

8	6	2	5	9	1	4	3	7
4	9	1	8	3	7	5	6	2
7	5	3	6	4	2	1	8	9
5	1	6	3	7	4	9	2	8
3	7	9	2	5	8	6	4	1
2	8	4	1	6	9	3	7	5
1	4	8	9	2	3	7	5	6
6	2	7	4	1	5	8	9	3
9	3	5	7	8	6	2	1	4

Expert Solution 152

8	4	7	3	5	1	2	6	9
6	1	5	7	9	2	8	4	3
2	9	3	4	6	8	7	5	1
3	7	4	9	1	5	6	2	8
9	5	6	8	2	3	1	7	4
1	8	2	6	7	4	9	3	5
7	6	1	5	4	9	3	8	2
5	3	9	2	8	6	4	1	7
4	2	8	1	3	7	5	9	6

Expert Solution 153

4	3	1	7	2	6	5	9	8
9	5	2	8	1	4	6	3	7
8	7	6	5	3	9	4	1	2
1	2	5	4	8	3	9	7	6
6	4	7	9	5	2	1	8	3
3	8	9	1	6	7	2	4	5
7	1	3	2	4	5	8	6	9
2	9	8	6	7	1	3	5	4
5	6	4	3	9	8	7	2	1

Expert Solution 154

7	1	2	5	6	4	9	3	8
6	3	4	1	8	9	7	2	5
8	5	9	3	2	7	4	1	6
2	4	8	9	5	3	1	6	7
5	9	1	6	7	2	8	4	3
3	7	6	4	1	8	2	5	9
4	8	5	2	9	6	3	7	1
1	2	7	8	3	5	6	9	4
9	6	3	7	4	1	5	8	2

Expert Solution 155

1	4	6	3	2	9	7	8	5
8	7	2	6	1	5	3	9	4
3	5	9	8	7	4	6	2	1
9	6	1	7	5	2	4	3	8
7	2	8	4	3	1	9	5	6
4	3	5	9	6	8	1	7	2
6	8	7	2	4	3	5	1	9
5	9	4	1	8	7	2	6	3
2	1	3	5	9	6	8	4	7

Expert Solution 156

6	7	3	1	2	5	4	8	9
9	1	4	8	6	3	2	5	7
2	5	8	7	4	9	1	3	6
7	2	9	5	8	4	3	6	1
5	4	1	6	3	2	9	7	8
8	3	6	9	7	1	5	2	4
4	6	5	3	9	7	8	1	2
1	8	2	4	5	6	7	9	3
3	9	7	2	1	8	6	4	5

Expert Solution 157

5	9	4	1	6	7	2	3	8
3	2	7	8	9	4	6	5	1
1	6	8	5	3	2	7	4	9
6	7	1	2	5	9	4	8	3
2	4	5	3	1	8	9	6	7
9	8	3	4	7	6	1	2	5
4	1	6	9	8	3	5	7	2
8	5	2	7	4	1	3	9	6
7	3	9	6	2	5	8	1	4

Expert Solution 158

5	4	7	9	1	2	6	3	8
6	1	9	8	5	3	4	7	2
8	2	3	4	7	6	5	9	1
3	9	4	6	8	1	2	5	7
7	8	1	2	9	5	3	4	6
2	6	5	3	4	7	8	1	9
1	7	8	5	6	4	9	2	3
9	5	2	7	3	8	1	6	4
4	3	6	1	2	9	7	8	5

Expert Solution 159

3	9	7	6	5	1	8	2	4
1	2	5	4	9	8	6	7	3
8	4	6	7	2	3	5	9	1
2	8	3	9	7	4	1	6	5
9	6	4	8	1	5	2	3	7
7	5	1	3	6	2	9	4	8
5	7	2	1	3	6	4	8	9
6	3	8	5	4	9	7	1	2
4	1	9	2	8	7	3	5	6

Expert Solution 160

5	4	2	3	1	6	7	8	9
9	6	3	7	2	8	4	5	1
7	1	8	4	5	9	3	6	2
2	7	4	6	3	5	1	9	8
3	8	9	2	4	1	6	7	5
6	5	1	9	8	7	2	4	3
1	9	6	8	7	3	5	2	4
4	3	7	5	9	2	8	1	6
8	2	5	1	6	4	9	3	7

Expert Solution 161

2	8	7	6	1	5	9	3	4
4	1	9	7	3	8	6	2	5
3	6	5	4	9	2	8	7	1
1	5	4	9	7	6	2	8	3
9	2	8	3	5	4	7	1	6
6	7	3	2	8	1	4	5	9
5	9	2	1	4	7	3	6	8
7	4	1	8	6	3	5	9	2
8	3	6	5	2	9	1	4	7

Expert Solution 162

1	9	5	6	2	8	7	3	4
3	2	6	7	5	4	9	8	1
8	7	4	1	3	9	6	2	5
2	4	3	8	7	5	1	9	6
7	5	1	2	9	6	3	4	8
9	6	8	4	1	3	5	7	2
4	3	7	5	8	1	2	6	9
5	8	2	9	6	7	4	1	3
6	1	9	3	4	2	8	5	7

Expert Solution 163

9	5	8	3	4	7	6	1	2
3	7	2	6	9	1	8	5	4
6	4	1	8	2	5	3	7	9
7	8	4	1	3	2	5	9	6
5	6	3	9	8	4	1	2	7
2	1	9	7	5	6	4	8	3
4	9	6	2	1	8	7	3	5
1	2	5	4	7	3	9	6	8
8	3	7	5	6	9	2	4	1

Expert Solution 164

7	8	9	6	4	2	1	5	3
5	3	6	9	7	1	2	8	4
4	2	1	8	5	3	7	9	6
8	9	2	4	1	6	3	7	5
3	6	5	2	8	7	9	4	1
1	7	4	3	9	5	6	2	8
2	1	8	5	3	9	4	6	7
6	5	7	1	2	4	8	3	9
9	4	3	7	6	8	5	1	2

Expert Solution 165

7	5	1	6	9	3	4	2	8
6	2	4	8	1	7	5	9	3
3	9	8	4	5	2	6	7	1
8	4	2	7	3	6	1	5	9
9	3	6	5	4	1	2	8	7
1	7	5	2	8	9	3	4	6
4	6	3	9	2	8	7	1	5
2	1	9	3	7	5	8	6	4
5	8	7	1	6	4	9	3	2

Expert Solution 166

5	9	2	6	3	8	1	7	4
4	6	3	5	7	1	9	2	8
1	8	7	2	9	4	6	3	5
9	4	8	3	1	2	7	5	6
2	1	5	7	6	9	4	8	3
7	3	6	8	4	5	2	1	9
6	5	1	9	2	3	8	4	7
8	7	4	1	5	6	3	9	2
3	2	9	4	8	7	5	6	1

Expert Solution 167

1	4	7	8	3	6	9	2	5
2	3	6	9	4	5	1	7	8
5	8	9	1	7	2	6	3	4
6	1	4	3	5	9	7	8	2
9	5	2	6	8	7	3	4	1
3	7	8	2	1	4	5	6	9
8	2	1	7	9	3	4	5	6
7	9	5	4	6	8	2	1	3
4	6	3	5	2	1	8	9	7

Expert Solution 168

5	9	7	4	8	6	2	3	1
6	4	8	1	3	2	9	5	7
3	1	2	9	7	5	6	4	8
2	6	1	5	9	3	7	8	4
7	3	5	2	4	8	1	6	9
4	8	9	6	1	7	3	2	5
1	5	4	3	6	9	8	7	2
9	7	6	8	2	4	5	1	3
8	2	3	7	5	1	4	9	6

Expert Solution 169

7	4	5	1	2	3	9	6	8
6	9	8	7	5	4	2	3	1
2	1	3	9	6	8	7	4	5
1	7	2	3	4	5	6	8	9
8	5	4	6	9	1	3	2	7
3	6	9	2	8	7	5	1	4
5	2	1	8	7	6	4	9	3
9	3	7	4	1	2	8	5	6
4	8	6	5	3	9	1	7	2

Expert Solution 170

4	6	1	7	2	9	5	3	8
3	2	7	8	5	1	9	4	6
5	9	8	4	6	3	7	2	1
1	4	5	2	9	8	6	7	3
7	8	6	5	3	4	1	9	2
9	3	2	6	1	7	8	5	4
8	5	4	1	7	2	3	6	9
2	7	9	3	8	6	4	1	5
6	1	3	9	4	5	2	8	7

Expert Solution 171

3	8	4	5	2	7	1	6	9
1	2	7	8	6	9	5	3	4
9	5	6	3	1	4	8	2	7
7	6	8	9	4	2	3	1	5
5	4	1	7	8	3	6	9	2
2	9	3	6	5	1	7	4	8
4	1	5	2	3	8	9	7	6
6	7	2	1	9	5	4	8	3
8	3	9	4	7	6	2	5	1

Expert Solution 172

5	8	4	1	6	2	9	3	7
1	2	7	3	9	8	6	5	4
9	3	6	5	4	7	2	1	8
3	5	8	4	7	9	1	6	2
2	4	1	6	8	3	7	9	5
7	6	9	2	1	5	8	4	3
8	7	3	9	5	6	4	2	1
6	1	2	7	3	4	5	8	9
4	9	5	8	2	1	3	7	6

Expert Solution 173

3	2	7	4	5	8	9	6	1
8	5	9	2	6	1	3	4	7
4	6	1	3	7	9	8	5	2
5	7	8	9	3	6	2	1	4
9	4	6	8	1	2	7	3	5
2	1	3	5	4	7	6	8	9
1	3	2	7	8	5	4	9	6
7	8	5	6	9	4	1	2	3
6	9	4	1	2	3	5	7	8

Expert Solution 174

3	1	8	9	4	7	5	6	2
4	6	5	2	3	8	7	9	1
7	9	2	6	1	5	8	4	3
6	2	9	1	8	4	3	5	7
1	5	7	3	6	2	4	8	9
8	3	4	5	7	9	1	2	6
5	7	6	4	9	1	2	3	8
9	4	1	8	2	3	6	7	5
2	8	3	7	5	6	9	1	4

Expert Solution 175

6	4	3	9	2	8	7	1	5
2	1	7	3	6	5	4	8	9
8	9	5	7	4	1	2	6	3
7	3	4	2	1	9	6	5	8
1	5	8	4	3	6	9	7	2
9	6	2	8	5	7	3	4	1
4	8	9	5	7	3	1	2	6
3	2	1	6	8	4	5	9	7
5	7	6	1	9	2	8	3	4

Expert Solution 176

7	2	3	1	5	8	9	6	4
1	4	5	9	7	6	8	2	3
6	9	8	4	3	2	7	5	1
3	7	2	8	6	9	1	4	5
5	6	4	2	1	7	3	9	8
8	1	9	3	4	5	2	7	6
4	8	7	5	2	3	6	1	9
9	5	6	7	8	1	4	3	2
2	3	1	6	9	4	5	8	7

Expert Solution 177

6	4	2	7	1	3	5	8	9
9	8	3	5	4	2	7	6	1
5	7	1	9	8	6	3	4	2
7	2	9	3	6	4	1	5	8
8	1	4	2	5	9	6	7	3
3	6	5	1	7	8	2	9	4
1	3	6	4	9	5	8	2	7
2	9	8	6	3	7	4	1	5
4	5	7	8	2	1	9	3	6

Expert Solution 178

8	7	2	6	5	3	9	4	1
3	4	5	9	2	1	7	6	8
1	6	9	7	8	4	3	2	5
4	9	6	2	1	7	8	5	3
5	3	7	8	4	6	1	9	2
2	1	8	5	3	9	6	7	4
9	8	4	3	6	2	5	1	7
6	2	3	1	7	5	4	8	9
7	5	1	4	9	8	2	3	6

Expert Solution 179

5	4	7	9	8	2	3	6	1
6	8	2	1	7	3	5	9	4
3	1	9	6	4	5	2	7	8
8	6	4	3	1	7	9	2	5
2	9	3	4	5	8	6	1	7
1	7	5	2	9	6	4	8	3
9	5	8	7	6	4	1	3	2
7	3	1	5	2	9	8	4	6
4	2	6	8	3	1	7	5	9

Expert Solution 180

7	2	8	6	4	1	9	3	5
5	3	6	7	2	9	4	1	8
4	9	1	3	8	5	2	7	6
6	1	2	9	7	4	5	8	3
8	5	4	2	3	6	1	9	7
9	7	3	1	5	8	6	4	2
3	4	9	5	6	7	8	2	1
1	6	7	8	9	2	3	5	4
2	8	5	4	1	3	7	6	9

Expert Solution 181

2	1	7	9	4	5	8	6	3
3	9	8	6	2	7	5	1	4
5	6	4	1	8	3	7	9	2
9	8	3	5	6	1	2	4	7
7	2	6	4	9	8	1	3	5
4	5	1	7	3	2	6	8	9
6	4	5	2	1	9	3	7	8
8	7	9	3	5	6	4	2	1
1	3	2	8	7	4	9	5	6

Expert Solution 182

3	7	8	9	2	5	6	4	1
4	5	9	3	1	6	7	8	2
1	6	2	8	7	4	3	5	9
8	4	1	2	6	7	5	9	3
6	9	7	5	3	1	8	2	4
5	2	3	4	9	8	1	6	7
2	8	6	7	4	3	9	1	5
7	1	4	6	5	9	2	3	8
9	3	5	1	8	2	4	7	6

Expert Solution 183

9	4	1	3	6	7	8	2	5
6	5	3	9	2	8	1	7	4
7	8	2	1	4	5	3	6	9
1	6	5	4	9	3	2	8	7
4	9	7	6	8	2	5	1	3
3	2	8	7	5	1	9	4	6
2	7	9	8	3	4	6	5	1
5	3	4	2	1	6	7	9	8
8	1	6	5	7	9	4	3	2

Expert Solution 184

3	9	5	7	2	1	4	6	8
7	6	8	4	9	5	2	3	1
1	4	2	8	3	6	7	5	9
6	1	3	5	8	2	9	4	7
4	5	7	3	1	9	8	2	6
2	8	9	6	4	7	5	1	3
9	3	1	2	7	4	6	8	5
5	7	4	1	6	8	3	9	2
8	2	6	9	5	3	1	7	4

Expert Solution 185

2	8	3	4	7	1	5	9	6
5	4	9	6	2	8	7	3	1
6	7	1	9	3	5	2	8	4
8	2	5	3	4	9	6	1	7
4	9	7	2	1	6	3	5	8
1	3	6	8	5	7	9	4	2
7	1	4	5	6	3	8	2	9
9	5	2	7	8	4	1	6	3
3	6	8	1	9	2	4	7	5

Expert Solution 186

6	2	5	8	7	1	9	4	3
4	9	7	5	3	2	8	1	6
1	8	3	9	4	6	5	2	7
8	7	6	2	5	9	4	3	1
9	5	1	3	6	4	7	8	2
3	4	2	7	1	8	6	9	5
5	6	9	4	2	3	1	7	8
7	3	4	1	8	5	2	6	9
2	1	8	6	9	7	3	5	4

Expert Solution 187

5	1	3	4	6	2	7	8	9
4	6	8	9	7	3	2	5	1
7	9	2	1	8	5	4	3	6
9	7	1	5	3	6	8	4	2
6	2	5	8	4	9	3	1	7
3	8	4	2	1	7	6	9	5
2	5	7	3	9	8	1	6	4
8	4	6	7	5	1	9	2	3
1	3	9	6	2	4	5	7	8

Expert Solution 188

6	5	8	2	7	1	9	3	4
3	4	9	8	6	5	2	7	1
2	7	1	9	3	4	8	5	6
7	3	2	1	8	6	4	9	5
8	1	4	5	9	7	3	6	2
5	9	6	3	4	2	7	1	8
4	6	3	7	5	8	1	2	9
9	2	5	4	1	3	6	8	7
1	8	7	6	2	9	5	4	3

Expert Solution 189

8	3	5	4	6	2	7	9	1
7	9	6	5	3	1	8	4	2
1	2	4	9	7	8	3	6	5
9	4	3	8	5	7	1	2	6
2	6	7	3	1	4	5	8	9
5	8	1	2	9	6	4	3	7
6	1	9	7	8	3	2	5	4
3	5	2	1	4	9	6	7	8
4	7	8	6	2	5	9	1	3

Expert Solution 190

5	2	3	7	1	6	8	9	4
6	7	4	8	5	9	1	3	2
9	1	8	3	4	2	7	5	6
2	4	7	5	9	1	3	6	8
3	5	6	4	7	8	9	2	1
8	9	1	2	6	3	4	7	5
1	6	2	9	3	4	5	8	7
7	8	9	1	2	5	6	4	3
4	3	5	6	8	7	2	1	9

Expert Solution 191

9	5	2	7	8	3	1	4	6
3	4	8	6	1	9	7	2	5
1	6	7	2	5	4	9	8	3
5	2	9	8	3	7	6	1	4
4	7	3	1	6	2	8	5	9
6	8	1	9	4	5	2	3	7
8	9	4	5	2	6	3	7	1
2	3	6	4	7	1	5	9	8
7	1	5	3	9	8	4	6	2

Expert Solution 192

9	8	5	3	4	6	1	7	2
7	6	1	5	9	2	4	3	8
3	2	4	8	7	1	9	6	5
4	5	3	7	8	9	6	2	1
6	9	8	2	1	3	7	5	4
2	1	7	4	6	5	3	8	9
5	7	2	1	3	4	8	9	6
8	4	6	9	2	7	5	1	3
1	3	9	6	5	8	2	4	7

Expert Solution 193

7	1	6	8	2	9	3	4	5
5	9	8	4	1	3	7	6	2
4	2	3	6	7	5	1	8	9
1	3	7	5	6	2	4	9	8
6	4	2	1	9	8	5	7	3
8	5	9	3	4	7	6	2	1
3	6	4	2	8	1	9	5	7
9	8	1	7	5	4	2	3	6
2	7	5	9	3	6	8	1	4

Expert Solution 194

1	5	4	6	8	9	3	7	2
9	6	2	7	1	3	4	5	8
7	8	3	4	5	2	1	6	9
4	9	8	1	3	6	5	2	7
6	2	1	5	9	7	8	4	3
5	3	7	8	2	4	9	1	6
3	4	9	2	6	1	7	8	5
2	7	5	3	4	8	6	9	1
8	1	6	9	7	5	2	3	4

Expert Solution 195

4	2	5	9	8	1	3	7	6
8	6	3	4	5	7	1	2	9
7	9	1	3	6	2	5	4	8
9	3	7	6	4	5	8	1	2
6	1	8	7	2	3	9	5	4
2	5	4	8	1	9	6	3	7
1	4	9	2	3	8	7	6	5
5	8	6	1	7	4	2	9	3
3	7	2	5	9	6	4	8	1

Expert Solution 196

9	3	1	2	4	8	5	7	6
5	7	8	9	6	1	2	3	4
6	2	4	5	3	7	1	8	9
4	6	3	8	1	5	7	9	2
8	1	9	7	2	4	3	6	5
7	5	2	6	9	3	8	4	1
1	4	7	3	5	6	9	2	8
3	9	5	4	8	2	6	1	7
2	8	6	1	7	9	4	5	3

Expert Solution 197

2	1	7	4	3	9	8	5	6
4	8	6	1	5	2	9	7	3
9	5	3	7	6	8	2	1	4
6	2	4	3	8	7	1	9	5
8	9	1	2	4	5	6	3	7
3	7	5	9	1	6	4	8	2
1	4	9	5	2	3	7	6	8
7	3	8	6	9	4	5	2	1
5	6	2	8	7	1	3	4	9

Expert Solution 198

5	9	1	7	2	4	6	3	8
8	7	2	6	9	3	5	1	4
4	6	3	5	1	8	9	2	7
3	2	5	9	6	7	4	8	1
9	1	8	4	3	2	7	5	6
6	4	7	8	5	1	2	9	3
1	3	9	2	7	6	8	4	5
2	8	6	3	4	5	1	7	9
7	5	4	1	8	9	3	6	2

Expert Solution 199

9	5	6	4	1	8	2	3	7
1	3	7	2	9	6	5	4	8
2	8	4	3	7	5	1	6	9
6	1	9	5	2	7	4	8	3
7	4	8	1	3	9	6	2	5
3	2	5	6	8	4	7	9	1
4	7	3	8	5	2	9	1	6
8	9	2	7	6	1	3	5	4
5	6	1	9	4	3	8	7	2

Expert Solution 200

4	9	8	2	3	6	7	5	1
3	6	1	7	5	9	8	4	2
2	5	7	4	1	8	9	3	6
1	2	6	8	4	3	5	7	9
8	4	9	5	2	7	6	1	3
7	3	5	9	6	1	4	2	8
9	1	4	6	7	2	3	8	5
6	7	2	3	8	5	1	9	4
5	8	3	1	9	4	2	6	7

Expert Solution 201

2	4	6	9	5	7	1	8	3
1	7	5	8	2	3	4	6	9
3	9	8	4	6	1	2	5	7
5	8	7	3	1	4	6	9	2
6	3	1	2	7	9	5	4	8
4	2	9	5	8	6	7	3	1
7	1	4	6	9	8	3	2	5
8	5	3	1	4	2	9	7	6
9	6	2	7	3	5	8	1	4

Expert Solution 202

2	7	4	8	6	3	9	5	1
8	3	9	5	4	1	7	2	6
6	5	1	2	7	9	8	4	3
4	6	7	3	2	8	1	9	5
3	2	5	9	1	7	6	8	4
1	9	8	4	5	6	3	7	2
9	4	3	6	8	2	5	1	7
5	1	6	7	9	4	2	3	8
7	8	2	1	3	5	4	6	9

Expert Solution 203

4	2	6	7	8	1	9	5	3
3	1	8	6	5	9	7	2	4
9	5	7	3	4	2	1	8	6
5	7	9	8	6	4	3	1	2
6	8	1	2	9	3	4	7	5
2	3	4	1	7	5	6	9	8
1	9	5	4	2	6	8	3	7
8	6	3	5	1	7	2	4	9
7	4	2	9	3	8	5	6	1

Expert Solution 204

3	8	5	2	1	9	7	4	6
7	1	9	6	3	4	8	2	5
2	6	4	7	5	8	3	9	1
1	9	7	8	2	3	5	6	4
6	2	8	9	4	5	1	3	7
5	4	3	1	7	6	9	8	2
4	7	6	3	8	1	2	5	9
8	5	1	4	9	2	6	7	3
9	3	2	5	6	7	4	1	8

Expert Solution 205

6	5	9	2	4	3	8	7	1
2	1	8	6	9	7	4	3	5
3	4	7	1	8	5	2	6	9
5	8	6	9	7	2	1	4	3
1	9	2	3	6	4	7	5	8
7	3	4	5	1	8	9	2	6
9	7	1	4	5	6	3	8	2
4	6	3	8	2	9	5	1	7
8	2	5	7	3	1	6	9	4

Expert Solution 206

8	3	7	5	9	6	4	1	2
4	6	2	3	8	1	5	7	9
9	1	5	2	7	4	3	6	8
5	9	8	4	6	7	2	3	1
7	4	1	9	2	3	8	5	6
3	2	6	1	5	8	7	9	4
1	5	4	6	3	2	9	8	7
2	7	9	8	1	5	6	4	3
6	8	3	7	4	9	1	2	5

Expert Solution 207

6	8	5	3	1	7	9	4	2
4	2	9	5	6	8	3	7	1
1	3	7	2	9	4	5	6	8
3	5	1	4	8	2	6	9	7
7	6	8	9	5	3	1	2	4
9	4	2	1	7	6	8	5	3
2	1	6	7	3	5	4	8	9
5	7	3	8	4	9	2	1	6
8	9	4	6	2	1	7	3	5

Expert Solution 208

4	7	2	3	9	5	8	6	1
3	1	9	8	6	4	2	7	5
5	6	8	7	2	1	3	9	4
6	9	4	1	5	3	7	2	8
7	2	5	6	4	8	9	1	3
1	8	3	9	7	2	4	5	6
8	3	7	2	1	6	5	4	9
2	5	1	4	3	9	6	8	7
9	4	6	5	8	7	1	3	2

Expert Solution 209

7	1	3	5	8	4	2	6	9
4	6	8	9	1	2	5	3	7
2	5	9	3	7	6	4	8	1
3	9	5	6	2	1	7	4	8
1	7	2	4	9	8	6	5	3
6	8	4	7	5	3	9	1	2
9	4	6	1	3	7	8	2	5
8	3	7	2	4	5	1	9	6
5	2	1	8	6	9	3	7	4

Expert Solution 210

5	1	9	4	2	7	8	6	3
8	4	2	3	6	9	1	7	5
3	7	6	1	5	8	9	4	2
1	3	7	2	8	6	4	5	9
9	5	8	7	3	4	2	1	6
6	2	4	5	9	1	3	8	7
7	9	3	8	4	5	6	2	1
2	8	1	6	7	3	5	9	4
4	6	5	9	1	2	7	3	8

Expert Solution 211

4	6	8	7	5	9	3	1	2
7	5	9	2	3	1	8	4	6
3	2	1	4	6	8	9	7	5
8	9	4	5	7	6	2	3	1
5	3	2	8	1	4	7	6	9
1	7	6	9	2	3	4	5	8
2	1	7	3	9	5	6	8	4
9	4	5	6	8	7	1	2	3
6	8	3	1	4	2	5	9	7

Expert Solution 212

7	6	4	5	9	3	1	8	2
3	9	8	6	1	2	5	4	7
5	2	1	8	7	4	3	6	9
9	8	3	1	5	7	6	2	4
6	4	7	3	2	8	9	1	5
2	1	5	9	4	6	8	7	3
1	3	2	4	8	9	7	5	6
8	7	9	2	6	5	4	3	1
4	5	6	7	3	1	2	9	8

Expert Solution 213

8	4	3	9	7	2	1	6	5
6	7	1	8	4	5	2	3	9
5	9	2	3	1	6	8	7	4
4	6	8	7	2	9	3	5	1
7	1	5	4	8	3	6	9	2
2	3	9	5	6	1	7	4	8
1	2	4	6	9	7	5	8	3
9	5	7	1	3	8	4	2	6
3	8	6	2	5	4	9	1	7

Expert Solution 214

6	1	2	9	4	3	7	8	5
3	8	5	6	1	7	2	4	9
9	7	4	5	8	2	6	3	1
4	9	7	8	3	1	5	2	6
8	5	6	4	2	9	3	1	7
2	3	1	7	6	5	4	9	8
1	2	9	3	7	6	8	5	4
7	4	3	1	5	8	9	6	2
5	6	8	2	9	4	1	7	3

Expert Solution 215

6	1	5	3	4	9	2	7	8
2	9	8	5	6	7	4	3	1
7	3	4	1	2	8	9	6	5
5	2	3	6	1	4	8	9	7
9	7	1	2	8	3	5	4	6
4	8	6	7	9	5	3	1	2
8	5	9	4	7	6	1	2	3
3	6	2	9	5	1	7	8	4
1	4	7	8	3	2	6	5	9

Expert Solution 216

9	4	5	8	1	6	2	7	3
2	7	8	5	9	3	1	6	4
1	3	6	2	7	4	5	8	9
6	5	7	9	4	1	8	3	2
4	2	1	7	3	8	6	9	5
8	9	3	6	5	2	7	4	1
5	8	2	4	6	9	3	1	7
7	1	9	3	8	5	4	2	6
3	6	4	1	2	7	9	5	8

Expert Solution 217

8	2	7	9	4	6	3	5	1
3	6	1	8	5	7	4	9	2
4	9	5	1	2	3	7	8	6
6	8	3	4	1	5	9	2	7
7	4	9	3	8	2	1	6	5
1	5	2	7	6	9	8	4	3
2	3	4	6	9	1	5	7	8
9	1	6	5	7	8	2	3	4
5	7	8	2	3	4	6	1	9

Expert Solution 218

9	6	4	1	7	2	8	5	3
2	8	1	5	3	4	6	7	9
5	3	7	8	6	9	4	2	1
7	2	9	6	1	8	5	3	4
4	5	6	7	9	3	2	1	8
3	1	8	2	4	5	7	9	6
8	4	5	9	2	1	3	6	7
6	9	3	4	5	7	1	8	2
1	7	2	3	8	6	9	4	5

Expert Solution 219

1	3	9	8	2	5	6	4	7
5	6	7	4	9	1	2	3	8
2	8	4	6	7	3	1	9	5
8	9	6	3	4	2	7	5	1
3	7	1	5	6	9	8	2	4
4	5	2	1	8	7	3	6	9
6	1	3	9	5	8	4	7	2
9	2	8	7	3	4	5	1	6
7	4	5	2	1	6	9	8	3

Expert Solution 220

5	6	9	8	7	3	1	4	2
1	2	7	6	5	4	3	9	8
8	3	4	2	1	9	5	7	6
7	8	3	9	4	5	2	6	1
9	5	2	3	6	1	7	8	4
4	1	6	7	2	8	9	5	3
3	7	8	4	9	2	6	1	5
2	9	1	5	8	6	4	3	7
6	4	5	1	3	7	8	2	9

Expert Solution 221

2	5	4	7	1	3	9	8	6
7	6	8	4	5	9	3	1	2
9	1	3	6	8	2	7	4	5
1	3	5	9	7	6	8	2	4
6	4	9	2	3	8	5	7	1
8	2	7	1	4	5	6	3	9
5	7	2	8	6	4	1	9	3
4	8	6	3	9	1	2	5	7
3	9	1	5	2	7	4	6	8

Expert Solution 222

1	5	7	9	8	3	6	2	4
6	9	2	4	7	1	5	8	3
8	4	3	5	2	6	1	7	9
4	3	1	6	9	8	7	5	2
7	8	5	2	3	4	9	6	1
9	2	6	7	1	5	4	3	8
2	1	9	3	5	7	8	4	6
5	6	8	1	4	2	3	9	7
3	7	4	8	6	9	2	1	5

Expert Solution 223

2	3	6	1	9	4	8	5	7
9	4	7	8	6	5	3	1	2
1	8	5	7	3	2	4	9	6
6	9	1	4	2	3	7	8	5
7	2	3	5	1	8	6	4	9
4	5	8	9	7	6	2	3	1
8	1	9	6	4	7	5	2	3
3	6	4	2	5	1	9	7	8
5	7	2	3	8	9	1	6	4

Expert Solution 224

8	4	7	9	6	3	1	2	5
5	1	9	2	7	8	4	6	3
2	3	6	4	5	1	7	9	8
4	5	8	6	9	7	2	3	1
1	6	2	5	3	4	8	7	9
7	9	3	1	8	2	6	5	4
3	8	1	7	2	5	9	4	6
9	2	5	8	4	6	3	1	7
6	7	4	3	1	9	5	8	2

Expert Solution 225

9	6	1	5	2	8	4	3	7
8	7	5	3	1	4	6	9	2
4	3	2	6	9	7	5	8	1
1	5	3	2	7	6	8	4	9
2	8	9	4	3	5	1	7	6
6	4	7	9	8	1	3	2	5
7	9	8	1	5	3	2	6	4
3	1	4	7	6	2	9	5	8
5	2	6	8	4	9	7	1	3

Expert Solution 226

1	8	7	9	2	3	4	5	6
2	4	9	1	6	5	7	3	8
5	6	3	4	7	8	2	1	9
3	7	2	6	1	4	8	9	5
8	9	5	7	3	2	1	6	4
4	1	6	8	5	9	3	2	7
9	2	4	3	8	6	5	7	1
7	3	8	5	9	1	6	4	2
6	5	1	2	4	7	9	8	3

Expert Solution 227

7	2	6	3	9	5	1	8	4
1	4	9	7	8	2	6	3	5
5	8	3	6	4	1	7	9	2
4	7	2	8	5	3	9	6	1
9	6	8	2	1	4	3	5	7
3	5	1	9	6	7	4	2	8
6	3	7	4	2	8	5	1	9
2	9	5	1	7	6	8	4	3
8	1	4	5	3	9	2	7	6

Expert Solution 228

8	9	6	7	5	1	3	4	2
2	1	4	6	3	9	5	8	7
7	3	5	2	4	8	9	6	1
6	4	8	1	2	5	7	3	9
3	5	7	9	8	6	1	2	4
1	2	9	3	7	4	6	5	8
4	6	2	5	1	7	8	9	3
5	7	3	8	9	2	4	1	6
9	8	1	4	6	3	2	7	5

Expert Solution 229

8	9	7	4	1	3	6	5	2
1	6	5	2	8	7	3	4	9
2	4	3	9	5	6	7	8	1
9	2	1	3	6	4	5	7	8
4	5	8	7	9	1	2	6	3
7	3	6	8	2	5	9	1	4
5	7	9	1	3	8	4	2	6
6	8	2	5	4	9	1	3	7
3	1	4	6	7	2	8	9	5

Expert Solution 230

8	2	9	4	3	7	6	1	5
1	5	7	6	9	2	3	8	4
6	3	4	8	5	1	2	7	9
5	1	6	7	2	3	9	4	8
4	7	2	9	1	8	5	3	6
3	9	8	5	6	4	7	2	1
2	6	1	3	8	5	4	9	7
7	8	5	2	4	9	1	6	3
9	4	3	1	7	6	8	5	2

Expert Solution 231

6	7	8	1	3	4	2	5	9
1	5	3	9	6	2	4	7	8
9	4	2	7	8	5	6	3	1
5	1	7	4	2	6	9	8	3
3	6	4	8	5	9	7	1	2
8	2	9	3	1	7	5	4	6
4	9	1	6	7	8	3	2	5
7	3	5	2	9	1	8	6	4
2	8	6	5	4	3	1	9	7

Expert Solution 232

1	8	7	5	3	4	2	9	6
5	9	6	1	7	2	4	3	8
4	2	3	9	6	8	7	5	1
9	3	1	7	8	6	5	2	4
2	7	5	4	1	9	6	8	3
8	6	4	3	2	5	9	1	7
6	1	8	2	5	7	3	4	9
7	5	9	8	4	3	1	6	2
3	4	2	6	9	1	8	7	5

Expert Solution 233

9	2	1	3	8	7	5	4	6
5	6	4	9	1	2	8	3	7
3	7	8	5	6	4	2	1	9
8	4	2	7	3	6	1	9	5
6	9	5	1	2	8	3	7	4
7	1	3	4	9	5	6	2	8
2	5	7	8	4	3	9	6	1
4	3	9	6	5	1	7	8	2
1	8	6	2	7	9	4	5	3

Expert Solution 234

7	2	5	6	4	3	9	8	1
6	8	9	1	5	7	2	3	4
3	1	4	2	8	9	5	7	6
9	7	1	8	3	4	6	5	2
2	4	8	7	6	5	3	1	9
5	6	3	9	1	2	7	4	8
8	5	7	4	2	6	1	9	3
4	3	2	5	9	1	8	6	7
1	9	6	3	7	8	4	2	5

Expert Solution 235

2	6	7	9	4	1	5	8	3
4	3	9	5	8	7	2	1	6
5	8	1	3	2	6	7	4	9
8	2	3	6	5	4	9	7	1
9	5	6	7	1	8	3	2	4
7	1	4	2	9	3	8	6	5
1	9	2	8	6	5	4	3	7
3	4	5	1	7	2	6	9	8
6	7	8	4	3	9	1	5	2

Expert Solution 236

2	3	8	7	6	5	1	9	4
1	7	9	2	8	4	3	5	6
4	5	6	9	3	1	2	8	7
3	9	4	6	1	2	8	7	5
8	6	2	4	5	7	9	3	1
7	1	5	3	9	8	6	4	2
9	4	7	1	2	3	5	6	8
6	8	1	5	4	9	7	2	3
5	2	3	8	7	6	4	1	9

Expert Solution 237

9	2	8	6	1	4	3	7	5
6	7	1	3	5	9	2	4	8
5	4	3	2	8	7	6	9	1
2	6	4	8	7	3	5	1	9
3	1	9	5	4	2	7	8	6
7	8	5	1	9	6	4	2	3
1	9	2	4	6	5	8	3	7
4	5	7	9	3	8	1	6	2
8	3	6	7	2	1	9	5	4

Expert Solution 238

3	2	9	8	5	7	1	6	4
6	8	5	1	3	4	7	2	9
4	7	1	2	6	9	8	5	3
5	4	7	9	8	1	6	3	2
1	9	3	7	2	6	5	4	8
2	6	8	3	4	5	9	7	1
9	5	6	4	1	3	2	8	7
8	1	4	6	7	2	3	9	5
7	3	2	5	9	8	4	1	6

Expert Solution 239

9	4	5	2	7	3	1	6	8
8	6	2	1	5	4	3	7	9
1	7	3	9	8	6	5	2	4
5	8	4	6	3	9	2	1	7
3	1	7	5	4	2	9	8	6
2	9	6	8	1	7	4	5	3
4	5	9	7	6	1	8	3	2
7	2	1	3	9	8	6	4	5
6	3	8	4	2	5	7	9	1

Expert Solution 240

2	3	5	9	1	6	7	4	8
6	4	9	5	7	8	2	1	3
8	7	1	3	2	4	9	5	6
5	8	3	6	9	1	4	7	2
7	9	4	2	3	5	6	8	1
1	6	2	4	8	7	5	3	9
3	5	6	1	4	9	8	2	7
4	1	8	7	6	2	3	9	5
9	2	7	8	5	3	1	6	4

Expert Solution 241

8	3	6	9	1	2	7	4	5
7	1	2	3	5	4	8	9	6
9	4	5	7	6	8	1	3	2
4	5	1	6	7	3	2	8	9
6	8	9	4	2	1	5	7	3
2	7	3	8	9	5	6	1	4
3	2	8	5	4	7	9	6	1
5	6	4	1	8	9	3	2	7
1	9	7	2	3	6	4	5	8

Expert Solution 242

1	9	6	5	3	4	2	8	7
2	8	3	6	1	7	5	9	4
5	7	4	9	2	8	1	3	6
6	3	2	1	4	5	8	7	9
4	1	7	2	8	9	6	5	3
9	5	8	7	6	3	4	2	1
7	6	1	3	5	2	9	4	8
8	2	9	4	7	6	3	1	5
3	4	5	8	9	1	7	6	2

Expert Solution 243

1	5	3	2	9	8	7	6	4
6	2	7	5	1	4	8	3	9
4	9	8	6	3	7	2	5	1
7	8	1	9	6	5	3	4	2
3	4	2	8	7	1	5	9	6
9	6	5	4	2	3	1	7	8
5	1	4	3	8	6	9	2	7
2	7	6	1	5	9	4	8	3
8	3	9	7	4	2	6	1	5

Expert Solution 244

9	2	4	7	3	5	6	8	1
1	6	8	4	9	2	3	7	5
3	5	7	8	1	6	2	4	9
7	3	1	9	6	8	5	2	4
5	8	6	3	2	4	1	9	7
4	9	2	1	5	7	8	6	3
2	4	3	5	8	9	7	1	6
8	7	5	6	4	1	9	3	2
6	1	9	2	7	3	4	5	8

Expert Solution 245

3	2	8	1	5	7	6	4	9
6	1	9	8	2	4	5	7	3
7	4	5	6	3	9	2	1	8
1	8	2	5	6	3	4	9	7
9	5	3	7	4	8	1	6	2
4	6	7	9	1	2	8	3	5
2	7	6	4	9	5	3	8	1
8	3	1	2	7	6	9	5	4
5	9	4	3	8	1	7	2	6

Expert Solution 246

3	1	9	7	8	4	6	2	5
4	5	8	2	1	6	7	9	3
2	6	7	3	9	5	4	8	1
7	9	3	1	5	8	2	4	6
6	4	2	9	7	3	5	1	8
5	8	1	6	4	2	3	7	9
8	3	4	5	2	1	9	6	7
9	2	5	8	6	7	1	3	4
1	7	6	4	3	9	8	5	2

Expert Solution 247

7	4	8	5	2	6	3	9	1
5	1	9	4	3	7	6	2	8
6	2	3	9	8	1	4	7	5
3	6	4	7	1	5	9	8	2
8	7	5	3	9	2	1	4	6
1	9	2	6	4	8	7	5	3
9	8	1	2	6	4	5	3	7
2	3	7	1	5	9	8	6	4
4	5	6	8	7	3	2	1	9

Expert Solution 248

5	3	6	2	1	7	4	8	9
1	2	8	4	6	9	3	5	7
4	9	7	3	8	5	6	2	1
3	8	9	6	5	2	7	1	4
7	1	2	9	3	4	8	6	5
6	4	5	8	7	1	2	9	3
8	5	3	1	4	6	9	7	2
2	7	4	5	9	8	1	3	6
9	6	1	7	2	3	5	4	8

Expert Solution 249

8	1	9	5	4	6	7	3	2
3	6	7	8	2	9	4	5	1
2	5	4	1	7	3	6	8	9
5	9	1	6	8	7	2	4	3
6	2	8	4	3	1	9	7	5
7	4	3	9	5	2	8	1	6
9	8	6	7	1	5	3	2	4
1	7	2	3	6	4	5	9	8
4	3	5	2	9	8	1	6	7

Expert Solution 250

3	5	6	9	1	2	8	4	7
9	2	7	8	6	4	3	5	1
1	8	4	3	5	7	9	6	2
5	7	9	2	4	3	1	8	6
8	6	3	1	7	9	5	2	4
2	4	1	6	8	5	7	3	9
6	9	8	4	3	1	2	7	5
7	3	2	5	9	6	4	1	8
4	1	5	7	2	8	6	9	3

Expert Solution 251

7	3	5	9	1	8	6	4	2
9	8	4	2	6	5	1	7	3
1	6	2	3	7	4	8	5	9
8	2	1	5	4	3	9	6	7
5	4	3	7	9	6	2	8	1
6	7	9	8	2	1	5	3	4
2	1	8	6	3	7	4	9	5
4	5	7	1	8	9	3	2	6
3	9	6	4	5	2	7	1	8

Expert Solution 252

4	1	9	3	6	8	7	5	2
7	8	2	9	5	1	3	6	4
5	3	6	4	2	7	8	1	9
6	7	8	2	1	9	5	4	3
9	4	3	5	8	6	1	2	7
2	5	1	7	4	3	6	9	8
3	6	4	1	7	2	9	8	5
1	9	5	8	3	4	2	7	6
8	2	7	6	9	5	4	3	1

Expert Solution 253

4	1	6	2	8	5	3	7	9
5	9	7	3	6	1	4	8	2
3	2	8	4	7	9	1	5	6
9	8	5	6	1	3	2	4	7
7	6	4	9	5	2	8	3	1
1	3	2	7	4	8	9	6	5
6	4	3	1	2	7	5	9	8
8	7	1	5	9	4	6	2	3
2	5	9	8	3	6	7	1	4

Expert Solution 254

6	3	1	5	8	9	2	7	4
9	4	8	6	7	2	3	5	1
2	7	5	1	4	3	6	8	9
7	8	2	3	9	4	1	6	5
5	6	4	7	2	1	9	3	8
3	1	9	8	5	6	4	2	7
1	5	3	4	6	7	8	9	2
8	2	6	9	1	5	7	4	3
4	9	7	2	3	8	5	1	6

Expert Solution 255

5	7	8	3	1	9	4	6	2
1	2	4	5	6	8	7	3	9
3	6	9	7	2	4	5	8	1
2	1	6	8	7	5	3	9	4
9	4	5	1	3	2	8	7	6
7	8	3	4	9	6	1	2	5
8	5	2	9	4	7	6	1	3
4	9	1	6	8	3	2	5	7
6	3	7	2	5	1	9	4	8

Expert Solution 256

4	6	1	7	2	5	9	3	8
2	9	3	1	8	6	5	4	7
7	5	8	9	4	3	6	2	1
1	8	4	5	6	9	3	7	2
6	7	2	4	3	8	1	5	9
9	3	5	2	1	7	8	6	4
5	1	9	6	7	4	2	8	3
3	4	6	8	9	2	7	1	5
8	2	7	3	5	1	4	9	6

Expert Solution 257

1	8	3	4	6	7	5	9	2
5	7	9	8	2	3	4	6	1
6	2	4	1	5	9	7	8	3
7	4	8	2	9	1	3	5	6
3	5	1	7	4	6	9	2	8
9	6	2	5	3	8	1	4	7
2	9	7	3	8	5	6	1	4
8	3	5	6	1	4	2	7	9
4	1	6	9	7	2	8	3	5

Expert Solution 258

5	1	9	6	3	2	7	4	8
3	2	7	5	4	8	1	9	6
6	8	4	7	9	1	5	2	3
9	3	2	8	7	6	4	1	5
1	7	8	4	2	5	6	3	9
4	6	5	3	1	9	8	7	2
2	5	1	9	8	4	3	6	7
7	4	6	2	5	3	9	8	1
8	9	3	1	6	7	2	5	4

Expert Solution 259

8	4	6	1	9	7	5	3	2
2	1	7	5	4	3	6	8	9
5	3	9	8	6	2	1	4	7
3	9	8	2	1	4	7	6	5
6	5	2	3	7	9	8	1	4
1	7	4	6	5	8	2	9	3
9	6	5	7	3	1	4	2	8
7	8	3	4	2	6	9	5	1
4	2	1	9	8	5	3	7	6

Expert Solution 260

6	1	5	4	2	9	8	7	3
8	4	3	5	7	1	2	9	6
7	9	2	3	8	6	1	5	4
1	5	6	9	3	4	7	2	8
4	2	8	7	6	5	3	1	9
9	3	7	2	1	8	4	6	5
3	7	4	6	9	2	5	8	1
2	6	1	8	5	3	9	4	7
5	8	9	1	4	7	6	3	2

Expert Solution 261

2	7	1	5	3	6	9	8	4
4	9	5	7	8	2	1	3	6
3	8	6	4	1	9	2	7	5
8	5	4	3	2	1	6	9	7
7	6	2	8	9	4	5	1	3
1	3	9	6	7	5	8	4	2
6	4	8	9	5	3	7	2	1
5	1	7	2	4	8	3	6	9
9	2	3	1	6	7	4	5	8

Expert Solution 262

3	8	1	5	4	9	7	6	2
4	6	2	1	8	7	5	3	9
7	5	9	2	6	3	8	1	4
1	2	7	3	9	5	4	8	6
8	9	3	4	1	6	2	7	5
6	4	5	7	2	8	3	9	1
9	3	4	8	5	1	6	2	7
2	7	6	9	3	4	1	5	8
5	1	8	6	7	2	9	4	3

Expert Solution 263

6	1	4	5	3	2	7	9	8
2	5	3	9	7	8	6	4	1
7	9	8	4	1	6	2	3	5
4	2	1	3	5	7	9	8	6
3	8	9	2	6	4	5	1	7
5	7	6	8	9	1	3	2	4
8	3	7	6	4	9	1	5	2
9	6	2	1	8	5	4	7	3
1	4	5	7	2	3	8	6	9

Expert Solution 264

4	8	3	7	2	5	9	6	1
2	1	7	9	4	6	8	3	5
5	9	6	8	1	3	4	2	7
3	4	1	2	8	7	5	9	6
9	7	5	3	6	4	1	8	2
8	6	2	5	9	1	7	4	3
7	2	4	1	3	9	6	5	8
6	5	8	4	7	2	3	1	9
1	3	9	6	5	8	2	7	4

Expert Solution 265

4	6	1	3	2	5	9	8	7
3	8	5	9	7	4	1	6	2
7	2	9	8	6	1	4	3	5
2	3	4	1	9	7	6	5	8
9	5	6	4	3	8	7	2	1
8	1	7	2	5	6	3	9	4
1	9	3	5	4	2	8	7	6
5	7	8	6	1	3	2	4	9
6	4	2	7	8	9	5	1	3

Expert Solution 266

1	3	5	4	6	2	9	8	7
4	7	8	1	9	5	3	2	6
2	6	9	3	7	8	1	5	4
7	8	4	6	2	9	5	1	3
6	9	2	5	1	3	7	4	8
3	5	1	7	8	4	6	9	2
5	2	6	8	3	1	4	7	9
8	4	7	9	5	6	2	3	1
9	1	3	2	4	7	8	6	5

Expert Solution 267

7	9	4	3	1	2	6	8	5
3	2	6	7	5	8	4	9	1
1	5	8	6	9	4	2	3	7
2	6	7	9	8	5	3	1	4
5	4	9	1	7	3	8	6	2
8	3	1	2	4	6	5	7	9
9	8	2	4	6	7	1	5	3
4	1	5	8	3	9	7	2	6
6	7	3	5	2	1	9	4	8

Expert Solution 268

4	5	3	7	6	8	9	1	2
1	8	2	5	4	9	3	6	7
6	9	7	1	3	2	8	4	5
5	6	4	9	2	7	1	8	3
2	1	8	6	5	3	4	7	9
7	3	9	8	1	4	2	5	6
8	2	6	3	7	1	5	9	4
9	4	5	2	8	6	7	3	1
3	7	1	4	9	5	6	2	8

Expert Solution 269

9	2	4	3	6	7	5	1	8
5	7	1	8	2	9	4	3	6
6	3	8	4	5	1	2	9	7
1	6	5	9	3	8	7	2	4
2	4	9	1	7	5	6	8	3
7	8	3	2	4	6	9	5	1
8	1	7	6	9	2	3	4	5
4	9	6	5	1	3	8	7	2
3	5	2	7	8	4	1	6	9

Expert Solution 270

7	6	3	8	9	4	5	1	2
9	5	8	1	6	2	7	4	3
1	4	2	3	5	7	8	9	6
4	7	9	5	8	6	3	2	1
2	3	5	4	1	9	6	8	7
8	1	6	7	2	3	9	5	4
5	2	7	9	3	1	4	6	8
3	9	1	6	4	8	2	7	5
6	8	4	2	7	5	1	3	9

Expert Solution 271

9	8	2	6	4	1	5	7	3
6	4	5	7	8	3	9	1	2
3	7	1	9	5	2	6	4	8
1	6	4	8	7	5	3	2	9
8	5	3	1	2	9	7	6	4
2	9	7	4	3	6	8	5	1
4	1	9	3	6	7	2	8	5
7	2	8	5	9	4	1	3	6
5	3	6	2	1	8	4	9	7

Expert Solution 272

1	2	4	6	3	9	5	8	7
5	6	8	7	1	2	9	4	3
7	3	9	8	5	4	6	1	2
4	7	5	9	2	1	8	3	6
8	9	2	4	6	3	1	7	5
6	1	3	5	8	7	4	2	9
3	5	7	1	4	6	2	9	8
9	4	6	2	7	8	3	5	1
2	8	1	3	9	5	7	6	4

Expert Solution 273

5	3	2	9	1	4	7	6	8
8	4	6	7	2	3	5	9	1
1	7	9	6	5	8	2	4	3
7	9	3	1	4	6	8	5	2
2	6	8	5	3	9	1	7	4
4	1	5	2	8	7	9	3	6
6	5	1	4	9	2	3	8	7
3	2	7	8	6	5	4	1	9
9	8	4	3	7	1	6	2	5

Expert Solution 274

7	3	8	2	1	4	5	6	9
2	9	4	7	6	5	8	1	3
6	1	5	8	3	9	2	4	7
5	2	9	3	4	1	7	8	6
8	7	3	6	5	2	4	9	1
4	6	1	9	7	8	3	2	5
1	5	6	4	8	3	9	7	2
3	4	2	1	9	7	6	5	8
9	8	7	5	2	6	1	3	4

Expert Solution 275

5	1	7	8	6	2	9	4	3
8	9	2	3	7	4	1	6	5
4	6	3	1	9	5	8	7	2
7	4	8	2	3	1	5	9	6
3	2	1	6	5	9	7	8	4
9	5	6	7	4	8	2	3	1
6	7	5	9	1	3	4	2	8
1	8	9	4	2	6	3	5	7
2	3	4	5	8	7	6	1	9

Expert Solution 276

8	1	2	4	3	6	7	5	9
3	5	9	2	8	7	1	6	4
7	4	6	5	1	9	2	3	8
4	8	1	3	7	5	6	9	2
9	7	5	6	2	1	8	4	3
6	2	3	9	4	8	5	1	7
5	3	8	7	6	4	9	2	1
1	9	4	8	5	2	3	7	6
2	6	7	1	9	3	4	8	5

Expert Solution 277

5	7	8	9	1	3	2	4	6
9	4	1	6	2	8	3	5	7
6	2	3	4	7	5	9	1	8
8	9	5	2	4	7	6	3	1
3	6	7	1	5	9	8	2	4
2	1	4	3	8	6	7	9	5
4	5	6	7	9	2	1	8	3
1	3	9	8	6	4	5	7	2
7	8	2	5	3	1	4	6	9

Expert Solution 278

6	4	1	3	5	8	9	2	7
3	7	2	1	4	9	5	8	6
5	8	9	6	7	2	1	3	4
7	5	6	9	2	4	8	1	3
2	1	8	7	3	5	4	6	9
4	9	3	8	1	6	7	5	2
8	3	4	5	6	7	2	9	1
9	6	7	2	8	1	3	4	5
1	2	5	4	9	3	6	7	8

Expert Solution 279

9	3	6	7	2	4	1	5	8
5	7	8	9	1	3	2	4	6
1	4	2	6	8	5	9	3	7
8	1	4	2	3	6	5	7	9
3	9	7	5	4	8	6	2	1
6	2	5	1	7	9	4	8	3
4	8	9	3	6	2	7	1	5
7	5	3	4	9	1	8	6	2
2	6	1	8	5	7	3	9	4

Expert Solution 280

5	3	2	9	6	4	8	7	1
4	8	7	5	1	3	2	9	6
6	1	9	7	2	8	4	3	5
2	5	8	6	9	1	7	4	3
1	6	3	4	8	7	9	5	2
7	9	4	3	5	2	1	6	8
3	4	1	8	7	6	5	2	9
8	7	5	2	3	9	6	1	4
9	2	6	1	4	5	3	8	7

Expert Solution 281

2	3	4	5	7	8	9	6	1
8	9	6	2	4	1	3	7	5
7	5	1	3	9	6	8	2	4
4	7	9	8	6	3	1	5	2
5	6	8	1	2	9	7	4	3
3	1	2	7	5	4	6	9	8
6	2	3	4	1	7	5	8	9
1	4	7	9	8	5	2	3	6
9	8	5	6	3	2	4	1	7

Expert Solution 282

8	1	3	2	9	6	5	7	4
5	7	2	3	8	4	1	9	6
4	6	9	7	5	1	8	2	3
6	2	5	9	3	7	4	1	8
1	9	8	6	4	5	7	3	2
3	4	7	1	2	8	9	6	5
9	8	1	4	6	2	3	5	7
2	3	4	5	7	9	6	8	1
7	5	6	8	1	3	2	4	9

Expert Solution 283

2	3	8	9	5	4	7	6	1
1	7	9	8	6	3	5	2	4
4	6	5	7	1	2	8	9	3
3	1	2	6	4	8	9	5	7
8	5	7	2	9	1	4	3	6
9	4	6	5	3	7	1	8	2
5	8	1	4	2	6	3	7	9
7	2	4	3	8	9	6	1	5
6	9	3	1	7	5	2	4	8

Expert Solution 284

1	8	5	6	4	7	9	3	2
9	7	2	1	5	3	8	4	6
4	3	6	2	9	8	7	1	5
8	9	3	7	2	6	4	5	1
2	1	4	5	3	9	6	7	8
5	6	7	4	8	1	2	9	3
6	5	1	9	7	2	3	8	4
7	2	8	3	1	4	5	6	9
3	4	9	8	6	5	1	2	7

Expert Solution 285

3	2	4	8	6	9	7	1	5
6	8	7	4	1	5	3	2	9
1	5	9	7	2	3	6	8	4
2	4	3	1	7	8	5	9	6
9	6	8	3	5	2	1	4	7
7	1	5	9	4	6	2	3	8
8	7	6	2	3	4	9	5	1
5	9	2	6	8	1	4	7	3
4	3	1	5	9	7	8	6	2

Expert Solution 286

3	9	2	6	7	5	4	1	8
6	7	1	2	8	4	3	9	5
8	5	4	3	1	9	6	2	7
1	4	5	7	2	8	9	6	3
7	3	6	4	9	1	8	5	2
2	8	9	5	6	3	7	4	1
9	6	3	8	5	2	1	7	4
4	2	7	1	3	6	5	8	9
5	1	8	9	4	7	2	3	6

Expert Solution 287

8	2	9	7	5	6	1	3	4
6	3	1	2	4	9	7	8	5
5	4	7	1	3	8	9	2	6
7	6	8	3	2	1	4	5	9
2	5	4	8	9	7	6	1	3
1	9	3	5	6	4	8	7	2
9	1	5	6	7	3	2	4	8
3	8	6	4	1	2	5	9	7
4	7	2	9	8	5	3	6	1

Expert Solution 288

8	7	5	9	4	2	6	1	3
1	2	3	6	5	8	7	4	9
9	6	4	3	1	7	8	2	5
6	5	1	4	2	3	9	7	8
2	4	7	8	9	5	1	3	6
3	8	9	1	7	6	2	5	4
5	9	6	2	3	1	4	8	7
7	1	8	5	6	4	3	9	2
4	3	2	7	8	9	5	6	1

Expert Solution 289

3	4	2	8	6	1	7	5	9
1	6	8	5	9	7	4	3	2
9	7	5	4	3	2	6	1	8
7	1	9	2	8	5	3	6	4
5	2	3	6	1	4	9	8	7
6	8	4	3	7	9	1	2	5
2	5	1	9	4	6	8	7	3
4	3	7	1	5	8	2	9	6
8	9	6	7	2	3	5	4	1

Expert Solution 290

6	3	8	7	4	1	5	9	2
4	7	5	9	6	2	8	1	3
2	1	9	8	5	3	4	7	6
3	6	7	5	9	8	1	2	4
8	9	1	6	2	4	3	5	7
5	4	2	3	1	7	6	8	9
9	8	3	1	7	6	2	4	5
7	2	6	4	8	5	9	3	1
1	5	4	2	3	9	7	6	8

Expert Solution 291

9	4	3	8	1	5	7	6	2
5	1	2	7	6	3	8	4	9
7	8	6	9	4	2	1	5	3
2	6	4	1	5	9	3	8	7
8	9	1	6	3	7	4	2	5
3	5	7	4	2	8	6	9	1
1	7	9	5	8	4	2	3	6
6	2	8	3	9	1	5	7	4
4	3	5	2	7	6	9	1	8

Expert Solution 292

4	9	7	3	6	8	1	5	2
8	1	3	2	5	9	6	4	7
6	2	5	4	7	1	9	3	8
3	7	2	6	8	5	4	9	1
1	6	4	9	2	3	7	8	5
5	8	9	7	1	4	3	2	6
2	3	1	8	9	7	5	6	4
7	4	8	5	3	6	2	1	9
9	5	6	1	4	2	8	7	3

Expert Solution 293

8	6	5	4	7	2	9	3	1
2	9	3	8	1	6	4	7	5
1	4	7	5	3	9	6	8	2
4	3	6	7	8	1	2	5	9
9	1	8	2	6	5	7	4	3
7	5	2	9	4	3	1	6	8
3	7	9	1	5	4	8	2	6
5	2	4	6	9	8	3	1	7
6	8	1	3	2	7	5	9	4

Expert Solution 294

8	6	1	5	7	3	2	4	9
5	3	2	9	4	8	7	6	1
7	9	4	1	2	6	3	8	5
1	5	6	2	3	4	8	9	7
2	4	9	8	5	7	6	1	3
3	7	8	6	9	1	4	5	2
6	1	3	7	8	9	5	2	4
9	2	7	4	6	5	1	3	8
4	8	5	3	1	2	9	7	6

Expert Solution 295

5	4	1	6	2	3	8	9	7
7	9	2	4	5	8	3	6	1
8	3	6	1	9	7	5	4	2
2	8	9	5	4	1	7	3	6
1	6	5	3	7	9	4	2	8
3	7	4	8	6	2	9	1	5
4	5	3	2	8	6	1	7	9
6	1	7	9	3	5	2	8	4
9	2	8	7	1	4	6	5	3

Expert Solution 296

2	9	8	4	7	5	3	6	1
6	5	7	1	8	3	4	9	2
4	1	3	9	6	2	8	5	7
8	2	5	7	4	6	9	1	3
9	3	4	8	5	1	2	7	6
7	6	1	3	2	9	5	4	8
1	8	6	2	9	4	7	3	5
3	7	9	5	1	8	6	2	4
5	4	2	6	3	7	1	8	9

Expert Solution 297

5	4	3	1	2	7	9	8	6
1	9	7	3	8	6	5	4	2
6	2	8	4	5	9	1	3	7
4	3	2	9	6	1	7	5	8
8	1	6	7	4	5	3	2	9
7	5	9	8	3	2	6	1	4
3	7	1	2	9	4	8	6	5
9	6	4	5	1	8	2	7	3
2	8	5	6	7	3	4	9	1

Expert Solution 298

8	1	4	9	5	3	2	7	6
2	3	6	7	1	4	5	8	9
7	5	9	8	2	6	3	4	1
1	9	3	2	4	5	7	6	8
5	8	2	6	9	7	4	1	3
6	4	7	1	3	8	9	2	5
3	7	1	5	8	2	6	9	4
9	6	5	4	7	1	8	3	2
4	2	8	3	6	9	1	5	7

Expert Solution 299

9	1	6	8	3	4	5	7	2
2	4	7	5	1	6	9	3	8
5	8	3	9	7	2	6	1	4
3	7	9	6	2	1	4	8	5
8	2	1	4	5	9	3	6	7
4	6	5	3	8	7	1	2	9
7	5	8	1	4	3	2	9	6
6	3	2	7	9	5	8	4	1
1	9	4	2	6	8	7	5	3

Expert Solution 300

6	1	4	5	9	2	8	3	7
5	3	7	8	1	6	9	2	4
9	8	2	4	3	7	6	1	5
8	6	1	3	5	9	7	4	2
2	9	3	7	4	1	5	8	6
7	4	5	6	2	8	3	9	1
3	7	9	1	6	4	2	5	8
4	2	8	9	7	5	1	6	3
1	5	6	2	8	3	4	7	9